Extended

# Cambridge IGCSE®
## Complete
# Mathematics
**Sixth Edition**

## Teacher Handbook

Ian Bettison
Mathew Taylor

Great Clarendon Street, Oxford, OX2 6DP, United Kingdom

Oxford University Press is a department of the University of Oxford. It furthers the University's objective of excellence in research, scholarship, and education by publishing worldwide. Oxford is a registered trade mark of Oxford University Press in the UK and in certain other countries.



British Library Cataloguing in Publication Data

Data available

9781382042543

10 9 8 7 6 5 4 3 2 1

Paper used in the production of this book is a natural, recyclable product made from wood grown in sustainable forests.

The manufacturing process conforms to the environmental regulations of the country of origin.

Printed in Great Britain by Ashford Colour Press Ltd

**Acknowledgements**

This Teacher Handbook refers to the Cambridge IGCSE® Mathematics (0580) syllabus published by Cambridge Assessment International Education.

This work has been developed independently from and is not endorsed by or otherwise connected with Cambridge Assessment International Education.

The publisher would like to thank the following for permissions to use copyright material:

**Cover:** Melinda Podor / Moment / Getty Images.

Artwork by Aptara Inc., Thompson Digital, and Oxford University Press.

Although we have made every effort to trace and contact all copyright holders before publication this has not been possible in all cases. If notified, the publisher will rectify any errors or omissions at the earliest opportunity.

# Contents


| | |
|---|---|
| **Introduction** | iv |
| **Content overview** | vi |

**1. Number 1** — 1
Investigations, activities and puzzles — 2
Glossary — 4
Extension worksheet — 6

**2. Algebra 1** — 7
Investigations, activities and puzzles — 8
Glossary — 10
Extension worksheet — 11

**3. Number 2** — 13
Investigations, activities and puzzles — 14
Glossary — 17
Extension worksheet — 18

**4. Trigonometry 1** — 19
Investigations, activities and puzzles — 20
Glossary — 23
Extension worksheet — 24

**5. Mensuration** — 25
Investigations, activities and puzzles — 26
Glossary — 28
Extension worksheet — 32

**6. Algebra 2** — 33
Investigations, activities and puzzles — 34
Glossary — 35
Extension worksheet — 36

**7. Geometry** — 37
Investigations, activities and puzzles — 38
Glossary — 40
Extension worksheet — 42

**8. Algebra 3** — 45
Investigations, activities and puzzles — 46
Glossary — 47
Extension worksheet — 48

**9. Graphs** — 49
Investigations, activities and puzzles — 50
Glossary — 52
Extension worksheet — 54

**10. Trigonometry 2** — 55
Investigations, activities and puzzles — 56
Glossary — 57
Extension worksheet — 58

**11. Sets and functions** — 59
Investigations, activities and puzzles — 60
Glossary — 62
Extension worksheet — 64

**12. Vectors and transformations** — 65
Investigations, activities and puzzles — 66
Glossary — 68
Extension worksheet — 69

**13. Statistics** — 71
Investigations, activities and puzzles — 72
Glossary — 74
Extension worksheet — 76

**14. Probability** — 77
Investigations, activities and puzzles — 78
Glossary — 80
Extension worksheet — 81

Revision checklists — 83
Multiple choice revision tests — 90
Examination-style Paper 2 (Non-calculator) — 98
Examination-style Paper 4 (Calculator) — 112
Answers and mark schemes — 125

# Introduction

Welcome to OUP's *Cambridge IGCSE® Complete Mathematics Extended: Teacher Handbook Sixth Edition.* This comprehensive guide is designed to support you as educators in delivering an enriching and effective learning experience for students pursuing the Cambridge IGCSE® in Mathematics.

As educators, you play a pivotal role in shaping the mathematical understanding and skills of your students. This handbook has been designed to provide you with practical activities and support resources that align with the *Cambridge IGCSE® Complete Mathematics Extended: Student Book Sixth Edition.* Whether you are a seasoned mathematics teacher or new to the Cambridge curriculum, this handbook is a valuable companion on your teaching journey.

## Content overview

Objectives for each chapter are listed and the relevant chapter(s) of both the Student Book and this handbook are mapped to each objective. This will enable you to quickly find the relevant teaching material, whatever programme of study you choose to adopt for delivering the material to your students.

## Chapter contents

Each chapter of this handbook is presented in the same format and with the same set of useful resources.

- **Chapter introduction** The introduction provides important information on the purpose of the chapter, a summary of the contents and prerequisite knowledge.
- **Investigations, activities and puzzles** These activities are designed to supplement the teaching of the syllabus content and provide students with opportunities for enrichment. Each activity has teaching notes to help you plan and deliver the activity as well as suggestions for extending the activities further.
- **Glossary** It is important that students understand the key terminology used in mathematics. These are listed, along with comprehensive definitions, in the glossary for each chapter.
- **Extension worksheet** More advanced students will make quicker progress through the contents of each chapter and we have provided an extension worksheet to enable these students to be stretched beyond the limits of the syllabus, while continuing to do relevant mathematics.

## Revision checklists

Each of the chapters in the book covers many aspects of the Cambridge IGCSE® syllabus. The revision checklists provide students with a summary list, in student-friendly language, to help them revise for the end-of-course examinations.

Each checklist can be copied and issued to the students, either at the end of the course or at an appropriate point during the course.

Students tick to indicate that they have revised the topics and whether they feel secure with their knowledge. This gives them responsibility for their learning and a clear idea of the progress they are making.

## Multiple choice revision tests

These tests are provided so that students can be quickly assessed on relevant content from across multiple chapters. It is recommended to do the tests at these points:

- Test 1 after Chapter 3
- Test 2 after Chapter 6
- Test 3 after Chapter 10
- Test 4 after Chapter 14

## Examination-style papers

There are two examination-style papers, one non-calculator and one calculator, that can be used either for revision or as a mock exam. They have been set to resemble a real examination paper and can be copied for students to complete on the paper, as they would do in the real exam.

## Answers and mark schemes

Comprehensive answers to the extension worksheets and multiple choice revision tests, as well as mark schemes for the examination-style papers, are included at the end of this handbook.

**kerboodle**

Kerboodle provides everything you need to plan, teach and track progress, including digital books, resources, assessment and automated Next Steps for each student.

We hope this handbook serves as a valuable resource, empowering you to inspire a love for mathematics and foster a deep understanding of the subject among your students.

*Ian Bettison*

*Mathew Taylor*

# Content overview

The Content overview shows you what topics and content are covered in each chapter. In some cases, the content splits into two or more discrete elements and is therefore covered in more than one chapter.

## Number

| Chapter in the Student Book | Topic |
|---|---|
| **Chapter 1** | Work with different types of numbers, e.g. natural numbers, integers, primes, squares, cubes, common factors, common multiples, rational numbers, irrational numbers, reciprocals. |
| **Chapter 1** | Calculate squares, square roots, cubes and cube roots and other powers and roots of numbers. |
| **Chapter 1** | Calculate with and convert between the following, including in contexts: proper fractions, improper fractions, mixed numbers, decimals and percentages. |
| **Chapter 1** | Order quantities and understand the symbols $=$, $\neq$, $>$, $<$, $\geqslant$, $\leqslant$. |
| **Chapter 1** | Calculate with integers, decimals and fractions including the correct order of operations. |
| **Chapter 1** | Understand and interpret positive, zero and negative indices including using the rules of indices. |
| **Chapter 1** | Understand and calculate with numbers in standard form $A \times 10^n$ where $n$ is a positive or negative integer, and $1 \leqslant A < 10$. |
| **Chapter 1** | Round to a given number of decimal places or significant figures and use rounded values to estimate calculations, including choosing an appropriate degree of accuracy when a question is in context. |
| **Chapter 1** | Understand and use upper and lower bounds including finding bounds of the results of calculations. |
| **Chapter 1** | Effective calculator use including entering values correctly and interpreting values correctly, e.g. for time and money. |
| **Chapter 1** | Understand and use surds, including in calculations, simplifying and rationalising the denominator. |
| **Chapter 3** | Calculate with and convert between the following, including in contexts: proper fractions, improper fractions, mixed numbers, decimals and percentages. |
| **Chapter 3** | Simplify a ratio, divide a quantity by a ratio. Use ratios in context. |
| **Chapter 3** | Understand and use measures of rate such as pressure and density and including solving problems involving average speed. |
| **Chapter 3** | Calculate a percentage of a quantity. Write one quantity as a percentage of another. Solve problems involving percentages including simple interest, compound interest, percentage increase/decrease and reverse percentages. |
| **Chapter 3** | Calculate with time including in the 24-hour and 12-hour clock. |
| **Chapter 3** | Calculate with money including converting between currencies. |
| **Chapter 9** | Use exponential growth and decay. |
| **Chapter 11** | Understand and use Venn diagrams and set notation. |

# Algebra and graphs

| Chapter in the Student Book | Topic |
| --- | --- |
| **Chapter 1** | Understand and use rules of indices including positive, zero and negative. |
| **Chapter 1** | Understand sequences including continuing a sequence, recognising patterns, term-to-term rules, finding and using $n$th terms and relationships between sequences. |
| **Chapter 2** | Substitute into expressions and formulae. |
| **Chapter 2** | Simplify expressions, expand brackets and factorise, including completing the square. |
| **Chapter 2** | Construct and solve linear equations including those where $x$ appears in the denominator as part of a linear expression. <br> Solve simultaneous equations, including where one is linear and one is nonlinear. <br> Solve quadratic equations including by factorising, completing the square and using the quadratic formula. <br> Change the subject of a formula. |
| **Chapter 6** | Simplify expressions, expand brackets and factorise, including completing the square. |
| **Chapter 6** | Construct and solve linear equations including those where $x$ appears in the denominator as part of a linear expression. <br> Solve simultaneous equations, including where one is linear and one is nonlinear. <br> Solve quadratic equations including by factorising, completing the square and using the quadratic formula. <br> Change the subject of a formula. |
| **Chapter 8** | Work with algebraic fractions including using the four rules, factorising and simplifying. |
| **Chapter 8** | Understand and use rules of indices including positive, zero and negative. |
| **Chapter 8** | Construct and solve linear equations including those where $x$ appears in the denominator as part of a linear expression. <br> Solve simultaneous equations, including where one is linear and one is nonlinear. <br> Solve quadratic equations including by factorising, completing the square and using the quadratic formula. <br> Change the subject of a formula. |
| **Chapter 8** | Understand and use linear inequalities including representing on a number line, constructing and solving and graphical representation of inequalities in two variables. |
| **Chapter 8** | Use and understand direct and inverse proportion including expressing algebraically and finding unknown quantitates. |
| **Chapter 9** | Use, understand and draw real-life graphs such as travel graphs and conversion graphs, distance–time graphs and speed–time graphs. |

| Chapter 9 | Draw, use and understand graphs including functions of the form $y = ax^n$ or $ab^x + c$ where $n$ is an integer between $-2$ and $3$ or $\pm\dfrac{1}{2}$, and graphs representing exponential growth and decay. <br><br> Solve associated equations graphically, e.g. find the intersection of a line and a curve |
|---|---|
| Chapter 9 | Sketch and interpret graphs of linear, quadratic, reciprocal and exponential graphs. |
| Chapter 9 | Understand differentiation including the derivatives of functions of the form $ax^n$ where $n$ is a positive integer or 0, gradients and stationary points, maximum and minimum points and estimating the gradient of a curve by drawing a tangent. |
| Chapter 11 | Functions including notation, domain, range, inverse functions and composite functions. |

## Coordinate geometry

| Chapter in the Student Book | Topic |
|---|---|
| Chapter 9 | Understand and use Cartesian coordinates. |
| Chapter 9 | Draw graphs of linear equations. |
| Chapter 9 | Find, use and interpret the gradient from a straight-line graph or from the coordinates of two points. |
| Chapter 9 | Find, use and interpret the length and midpoint of a line segment. |
| Chapter 9 | Find, use and interpret the equation of a straight-line graph. |
| Chapter 9 | Find, use and interpret the gradient and equation of a parallel line. |
| Chapter 9 | Find, use and interpret the gradient and equation of a perpendicular line. |

## Geometry

| Chapter in the Student Book | Topic |
|---|---|
| Chapter 4 | Understand and draw scale drawings including interpreting three-figure bearings. |
| Chapter 5 | Understand and use the ratio between lengths, areas and volumes of similar shapes and solids including solving problems. |
| Chapter 7 | Use and interpret geometrical terms and vocabulary. |
| Chapter 7 | Geometrical drawings and constructions including triangles and nets. |
| Chapter 7 | Symmetry in 2D and 3D shapes including line symmetry, order of rotational symmetry and axes of symmetry. |

| Chapter 7 | Use and understand geometrical properties including: |
|---|---|
| | • angles at a point |
| | • angles on a straight line |
| | • vertically opposite angles |
| | • sum of angles in a triangle |
| | • sum of angles in a quadrilateral |
| | • angles in parallel lines (corresponding, alternate, co-interior) |
| | • angle properties of regular and irregular polygons. |
| Chapter 7 | Understand and use geometrical properties of circles including: |
| | • angle between tangent and radius |
| | • angle in a semi-circle |
| | • angle at the centre and angle at the circumference |
| | • opposite angles in a cyclic quadrilateral |
| | • angles in the same segment |
| | • alternate segment theorem. |
| Chapter 7 | Understand and use geometrical properties of circles including: |
| | • equal chords are equal distances from the centre |
| | • the perpendicular bisector of a chord passes through the centre of the circle |
| | • tangents from an external point are equal. |

# Mensuration

| Chapter in the Student Book | Topic |
|---|---|
| Chapter 5 | Use and convert between units of mass, length, area, volume and capacity including in solving problems |
| Chapter 5 | Solve problems involving area and perimeter of triangles, rectangles, parallelograms and trapeziums. |
| Chapter 5 | Solve problems involving circumference and area of circles, arc length and sector area. |
| Chapter 5 | Solve problems involving surface area and volume of cuboids, prisms, cylinders, spheres, pyramids and cones. |
| Chapter 5 | Solve problems involving area and perimeter of compound shapes, including compound solids. |

# Trigonometry

| Chapter in the Student Book | Topic |
| --- | --- |
| **Chapter 4** | Calculate unknown sides of a right-angled triangle using Pythagoras' theorem, including solving problems in two dimensions. |
| **Chapter 4** | Calculate unknown sides and angles of a right-angled triangle using trigonometry, including solving problems in two dimensions and angles of elevation and depression. |
| **Chapter 4** | Know exact trigonometric values for $x = 0°$, $30°$, $45°$ and $60°$ (and $90°$ where applicable) |
| **Chapter 4** | Solve problems in three dimensions using Pythagoras' theorem and trigonometry. |
| **Chapter 5** | Use the sine rule and the cosine rule to find lengths and angles for non-right-angled triangles. |
| **Chapter 5** | Find the area of a triangle using $\frac{1}{2}ab\sin C$. |
| **Chapter 10** | Sketch and interpret the graphs of $\sin x$, $\cos x$ and $\tan x$. Solve trigonometric equations. |
| **Chapter 10** | Find the area of a triangle using $\frac{1}{2}ab\sin C$. |

# Transformations and vectors

| Chapter in the Student Book | Topic |
| --- | --- |
| **Chapter 12** | Draw and describe reflections, rotations, enlargements and translations. |
| **Chapter 12** | Understand and use vectors including describing a translation, adding and subtracting vectors and multiplying a vector by a scalar. |
| **Chapter 12** | Find the magnitude of a vector. |
| **Chapter 12** | Understand and use vector geometry including representing vectors by directed line segments, position vectors and vector geometry problems including using the sum and difference of vectors to express given vectors in terms of other vectors. |

## Probability

| Chapter in the Student Book | Topic |
|---|---|
| **Chapter 14** | Understand and use probability including the language, notation and the scale from 0 to 1. Calculate the probability of an event and understand the probability of an event not happening is 1 – (the probability of the event happening). |
| **Chapter 14** | Understand and use relative frequency to estimate probability and find expected frequencies. |
| **Chapter 14** | Find the probability of combined events using these diagrams: sample space diagrams, tables, Venn diagrams and tree diagrams. |
| **Chapter 14** | Calculate conditional probability using Venn diagrams, tree diagrams and tables. |

## Statistics

| Chapter in the Student Book | Topic |
|---|---|
| **Chapter 13** | Read, interpret, draw inferences from and compare data using tables, graphs and statistical diagrams and measures, including an appreciation of limits on any conclusion made. |
| **Chapter 13** | Find the mean, median, mode and range and choose the most appropriate statistical measures. Calculate an estimate of the mean and find the modal class from grouped data. |
| **Chapter 13** | Use tables for statistical data. |
| **Chapter 13** | Draw and interpret the following statistical diagrams: bar charts; pie charts; pictograms; stem-and-leaf diagrams; simple frequency distributions. |
| **Chapter 13** | Draw and interpret scatter diagrams (including understanding correlation and drawing a line of best fit). |
| **Chapter 13** | Draw and interpret cumulative frequency diagrams (including the median, percentiles, quartiles and interquartile range). |
| **Chapter 13** | Draw and interpret histograms (including understanding and using frequency density). |

The purpose of this first chapter is to make sure that students are familiar with numbers of various kinds, so that when they appear in other topics students will know how to use them.

The number types included in this chapter are:

- positive integers
- negative integers
- decimals
- proper and improper fractions
- powers of 10
- rational and irrational numbers
- surds
- numbers in standard form.

Students will also study these other skills:

- finding the nth term of a linear sequence
- finding the nth term of a quadratic sequence
- briefly considering cubic and exponential sequences
- working with upper and lower bounds
- estimating
- calculating with time and money.

There is less emphasis on working without a calculator in this chapter than in the Core course. There is a section near the end of the chapter designed to make sure that students are fully able to use a calculator to perform complex calculations when required.

After studying this chapter, students should be ready to use these numbers and skills in other contexts.

## Prerequisite knowledge

There is very little prerequisite knowledge required for studying this chapter.

Students should be able to count, know their basic multiplication tables, and be familiar with basic place value.

# Investigations, activities and puzzles

## Cutting paper

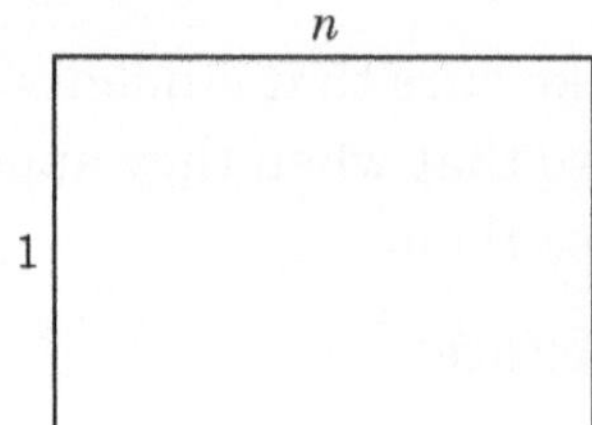


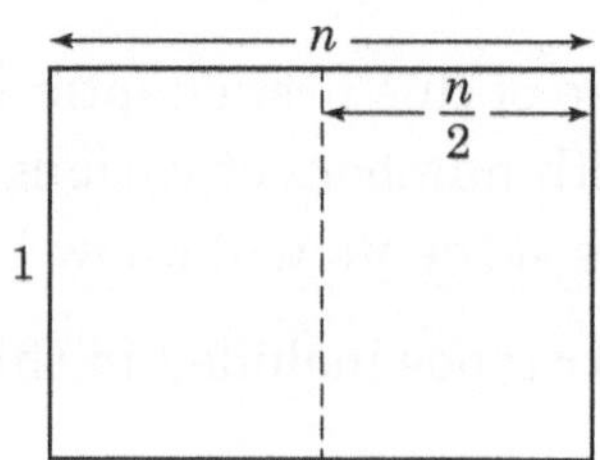


The sides of a rectangle are in the ratio $1:n$, but the value of $n$ is chosen so that when the rectangle is folded in half, the lengths of its sides are still in the ratio $1:n$.

In other words, the new rectangle is mathematically similar to the original rectangle.

What must the value of $n$ be for the rectangle to have this property?

Why is this a useful shape for paper used in business?

### Notes on this activity

This activity is useful for introducing the idea of surds, since the answer to the question is a rectangle whose sides are in the ratio $1:1\sqrt{2}$.

It would probably be a good idea for you to demonstrate to students what this activity is about by folding a piece of A4 paper in half in front of them, so that they can see that the folded paper has the same shape as the original sheet. You will probably find that very few students have realised that A4 paper has this property. It is clearly a useful property since it means that an A4 sheet can be enlarged or reduced easily, without changing its shape.

You can find a solution to the problem in the following way. If the folded rectangle is similar to the original rectangle, then

$$\frac{n}{1} = \frac{1}{\frac{n}{2}}$$

Cross multiplying gives

$$\frac{n^2}{2} = 1$$

and therefore

$$n^2 = 2 \quad \text{and} \quad n = \sqrt{2}$$

### Extension

Other questions you could consider are:

- Where else can you find the square root of 2 in mathematics?

- Can you prove that the square root of 2 is irrational?

## The biggest number

A particular calculator has the following buttons:

The only digit buttons that work on this calculator are '1', '2' and '3'.

a) You can press any button, but only once.
   What is the biggest number you can get?

b) Now the '1', '2', '3' and '4' buttons are working.
   What is the biggest number you can get?

c) Investigate what happens as you increase the number of digits that
   you can use.

**Notes on this activity**

This activity is really just a bit of fun to give students an
opportunity to experiment with the buttons on their calculators.

**Extension**

There are many ways in which this activity could be extended,
such as:

- give the students a target number to try to make
- allow them to use more functions on their calculators.

# Glossary

## Cube root

The cube root of a number is another number which, when cubed, will equal the first number. The cube root of 1 is 1.

### Example

The cube root of 8 is 2 since $2 \times 2 \times 2 = 8$

$\sqrt[3]{\ }$ is the symbol meaning 'find the cube root of the number given'.

## Decimal fraction

A decimal fraction is a way of expressing numbers less than 1 using the decimal place value system, extended to the right of the ones column so as to give values of $\frac{1}{10}$, $\frac{1}{100}$, and so on.

### Example

$0.376$ means $\frac{3}{10} + \frac{7}{100} + \frac{6}{1000} = \frac{376}{1000}$

## Estimate

An estimate is an approximation of a quantity made by judgement rather than by carrying out the process needed to produce an accurate answer. The process might be measuring, doing a sum or counting.

### Examples

An estimate of the number of people in a room might be 30, when the actual number found by counting is 27.

An estimate of $(23.7 \times 19.1) \div 99.6$ might be 4 or 5 or 4.5.

## Factor

A number that divides exactly into another number is called one of its factors.

### Example

3 is a factor of 24 because $24 \div 3 = 8$.

## Highest common factor

The highest common factor of two numbers is the highest number that is a factor of both of those numbers.

### Example

The factors of 60 are 1, 2, 3, 4, 5, 6, 10, 12, 15, 20, 30 and 60.

The factors of 66 are 1, 2, 3, 6, 11, 22, 33 and 66.

The highest number in both lists is 6, so 6 is the highest common factor of 60 and 66.

## Fraction

A fraction is a measure of how something is to be divided up or shared out. There are four principal ways of expressing fractions: common, decimal, percentage and ratio.

## Improper fraction

An improper fraction is a common fraction in which the numerator (the top number) is equal to or bigger than the denominator (the bottom number).

## Index or power

An index or power is a number or symbol written as a superscript to another number or symbol. When the index is a positive whole number, it indicates how many of the first number or symbol must be multiplied together. When the index is a fraction, then it indicates that a root has to be found.

### Examples

$A^2 = A \times A \quad 5^3 = 5 \times 5 \times 5 = 125 \quad 9^{\frac{1}{2}} = \sqrt{9} = 3$

## Integer

An integer is a number made from the natural numbers (including 0) by putting a positive or a negative sign in front. The positive sign is often omitted.

### Examples

$..., -5, -4, -3, -2, -1, 0, 1, 2, 3, 4, 5, ...$

## Irrational number

An irrational number can only be written (using no symbols) as a never-ending, non-repeating decimal fraction. An irrational number cannot be written in the form of a rational number.

### Examples

1.234567891011121314151617181920212223324...

The square root of any prime number is irrational, as is $\pi$.

## Lowest common multiple

The lowest common multiple of two numbers is the lowest number that is a multiple of both of those numbers.

**Examples**

The multiples of 3 are

3, 6, 9, 12, 15, 18, 21, 24, 27, ...

The multiples of 4 are

4, 8, 12, 16, 20, 24, 28, ...

The lowest number in both lists is 12, so 12 is the lowest common multiple of 3 and 4.

## Mixed number

A mixed number is made up of two parts: a whole number followed by a proper fraction.

**Examples**

$1\frac{1}{2}, 5\frac{7}{8}, -2\frac{5}{6}$ are all mixed numbers.

## Multiple

When you repeatedly add a number to itself you produce multiples of the original number.

**Example**

Starting with 7 and adding 7 repeatedly gives 7, 14, 21, 28, 35, etc. which are multiples of 7.

## Natural number

A natural number is a number in the set of numbers 1, 2, 3, 4, 5, 6, ... as used in counting. It is a matter of choice whether 0 is included or not.

## Negative number

A number less than zero.

## Positive number

A number greater than zero.

## Prime factor

A prime number which is a factor of another number one of its prime factors.

**Example**

Any positive number greater than 1 that is not prime can be expressed as a product of prime factors in only one way.

$30 = 2 \times 3 \times 5$

So, 3 is a factor of 30.

As 3 is prime, it is one of the prime factors of 30.

## Prime number

A prime number is a number having two, and only two, factors. Note that 1 is not a prime number since it has only one factor.

## Rational number

A rational number is a number that can be written in the form $\frac{a}{b}$ where $a$ and $b$ are integers and $b$ is not zero.

**Examples**

$-4.5 \quad 1\frac{1}{3} \quad 0.90909090909... \quad 3.\dot{1}4285\dot{7} \quad 8$

are all rational since they can be re-written as:

$\frac{-9}{2} \quad \frac{4}{3} \quad \frac{1}{11} \quad \frac{22}{7} \quad \frac{8}{1}$

## Reciprocal

The reciprocal of a number is the result of dividing 1 by that number.

For any whole number $n$, its reciprocal is $\frac{1}{n}$.

If the original number is a fraction, you can find its reciprocal by turning the fraction upside down.

**Examples**

The reciprocal of 2 is $\frac{1}{2}$.

The reciprocal of $\frac{2}{3}$ is $\frac{3}{2}$.

## Sequence

A sequence is a set of numbers or objects made and written in order according to some mathematical rule.

## Significant figures (s.f.)

Significant figures are used to express the relative importance of the digits in a number, the most important being the first digit from the left-hand end of the number that is not zero. Starting with the first non-zero digit, all digits are then counted as significant up to the last non-zero digit. After that, zeros may or may not be significant – it depends on the context.

**Examples**

3 s.f.   1230   78 500   16.3   0.0247   0.005 03

2 s.f.   1200   79 000   16   0.025   0.0050

## Square root

A square root of a number is another number which when squared will equal the first number. The square roots of 1 are 1 and −1.

**Example**

One square root of 16 is 4 because $4 \times 4 = 16$; another is −4 because $-4 \times -4 = 16$.

$\sqrt{\ }$ is the symbol meaning 'the square root of the number given'.

# Extension worksheet

This worksheet develops the work on sequences to include a formal treatment of arithmetic sequences.

1. How would you add up the numbers from 1 to 10?
   What about 1 to 100? What about 1 to 1000?

2. Find the following sums.

   **a)** $2 + 5 + 8 + 11$

   **b)** $1 + 3 + 5 + 7 + 9$

   **c)** $3 + 7 + 11 + 15 + 19$

   **d)** $4 + 3 + 2 + 1 + 0 + (-1)$

   **e)** $1.3 + 1.7 + 2.1 + 2.5$

   **f)** $6.5 + 7.2 + 7.9 + 8.6 + 9.3 + 10$

In all of the sums above the numbers form an *arithmetic* sequence – there is a common difference between the terms. Karl Friedrich Gauss devised a method for finding the sums of the terms of arithmetic sequences when he was at school and his teacher posed the problem of finding the sum of the numbers from 1 to 100. His method led to the general formula:

$$S_n = \frac{n}{2}(2a + (n-1)d)$$

where $S_n$ is the sum of $n$ terms in the sequence, $a$ is the first term and $d$ is the common difference.

Using the example of the sum of the numbers from 1 to 100, the formula gives:

$$S_{100} = \frac{100}{2}(2 \times 1 + (100 - 1) \times 1) = 5050$$

3. Use the formula to confirm your answers to the sums you found in Question **2**.

4. Find the following sums using the formula.

   **a)** $2 + 6 + 10 + 14 + 18 + 22 + 26 + 30 + 34 + 38 + 42 + 46 + 50$

   **b)** $1 + 3 + 5 + 7 + 9 + 11 + 13 + 15 + 17 + 19 + 21 + 23 + 25$

   **c)** The sum of the first 50 even numbers.

   **d)** The sum of the first 100 multiples of three.

   **e)** $1.3 + 1.6 + \cdots + 4.6 + 4.9$

5. Research the method used by Gauss to find the sum of the numbers from 1 to 100.

6. Make up some of your own arithmetic sequences and challenge a partner to find the sum of the terms.

(Answers on page 125)

# 2 Algebra 1

This chapter begins with a section on algebraic substitution. A basic familiarity with algebra is assumed, and the substitutions become increasingly more complicated as the questions progress.

The chapter then moves the focus to collecting 'like terms'. Again, these questions become increasingly difficult. Basic techniques such as expanding brackets are explained, but this will probably not be the first time that students have seen this.

Students are then required to expand two brackets and then three brackets, collecting like terms as they go.

The next section of this chapter requires students to solve equations. Some of the questions are quite demanding, and students are expected to be confident working with brackets, negative numbers and fractions. There is also an exercise of problems where students need to form their own equations before solving them.

The chapter closes with a section on simultaneous linear equations. The substitution and elimination methods are both discussed. There is also an exercise of problems where students need to form their own simultaneous equations before solving them.

## Prerequisite knowledge

Before attempting this chapter, students should be:

- familiar with the rules of BIDMAS
- able to work confidently with negative numbers
- able to work with fractions.

# Investigations, activities and puzzles

## Scales

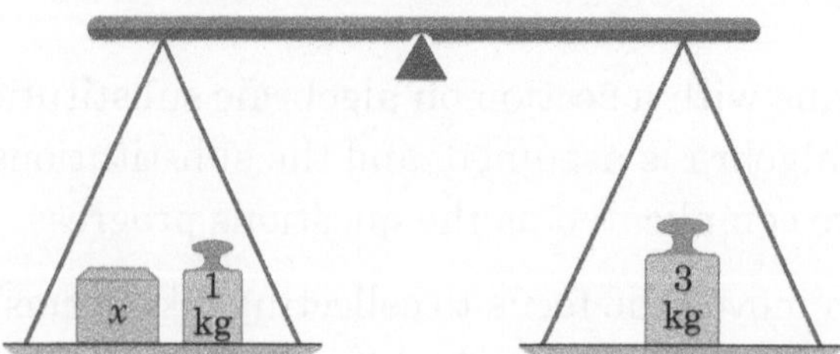


The diagram shows how the mass of the package $x$ can be measured using two masses.

If the scales are balanced, $x$ must be 2 kg.

1.  How can you measure all the masses from 1 kg to 10 kg using three masses: 1 kg, 3 kg, 6 kg?

2.  It is possible to measure all the masses from 1 kg to 13 kg using a different set of three masses. What are the three masses?

3.  It is possible to measure all the masses from 1 kg to 40 kg using four masses. What are the masses?

**Notes on this activity**

In terms of algebra content, the main ideas in this activity are that $x$ stands for a number and that this number can change.

It would be a good idea to discuss the first question in this activity with students. How would they do this for a randomly chosen value of $x$ within the range being considered?

Once Question **1** has been answered by the class, students can move on to Question **2**, which focuses on how the three values for the masses are chosen. How did they choose them? Was it simply trial and improvement or did they have any strategies?

Question **3** extends the problem so that they are using more masses to measure a greater number of kilograms.

## Extension

There are many other questions that could be considered for this problem. To measure all the masses from 1 kg to $n$ kg, students could consider some of these questions:

*   What is the greatest number of kilograms that can be measured using three masses? Why does this limit exist?
*   What is the greatest number of kilograms that can be measured using four masses? (This is an extension of the previous question.)
*   At what point would you need to introduce a fifth mass?
*   How do you choose which masses to use?
*   Are there some combinations of masses that are particularly bad choices?

## Simultaneous shapes

In the diagram, each shape is worth a particular value.
The row totals are given.

What is the star worth?

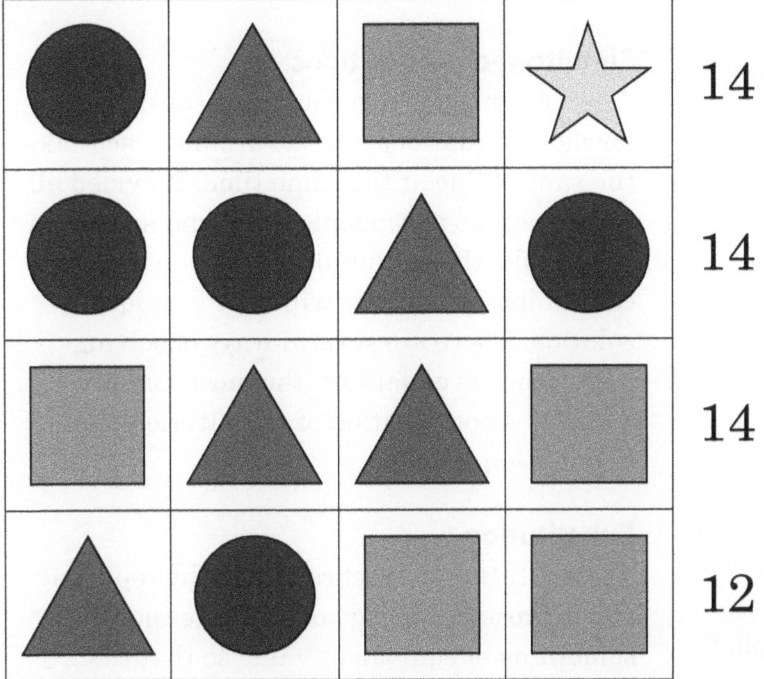

**Notes on this activity**

There are four unknowns, so students need to set up and solve four simultaneous equations.

They should find that the square is worth 2, the circle is worth 3, the triangle is worth 5 and the star is worth 4.

These kinds of puzzles are very popular as they provide students with a fun application of simultaneous equations.

Things you could consider with your students are:

- Can they make their own puzzles of a similar nature?
- What is the maximum number of different shapes they can use in a given size grid for the puzzle to be solvable?
- How does this relate to the solvability of equations with several variables?

# Glossary

### Equation

An equation is a statement that two expressions (one of which may be a constant) have the same value.

**Examples**

$2x + 7 = 15 \qquad 3(x + 5) = 3x + 15$

### Expression

An expression in algebra is a collection of terms, made up of constants and variables, linked by signs for operations and does not include an equals sign.

**Examples**

$x + y \quad 3 + x^2 - y \quad 4(x - y) \quad 3x^2 + 5y$

### Formula

A formula is a statement, usually written as an equation, giving the exact relationship between certain quantities, so that, when one or more values are known, the value of one particular quantity can be found.

**Example**

The volume $V$ of a sphere of diameter $d$ can be found using the formula $V = \dfrac{1}{6}\pi d^3$

### Linear equation

A linear equation is an equation involving only an expression, or expressions, of degree 1. Such an equation can be represented graphically by a straight line.

**Examples**

$y = 3x + 2 \quad y = 4 \quad x = 3y - 5$

### Sequence

A sequence is a set of numbers or objects made and written in order according to some mathematical rule.

### Simultaneous equations

A set of simultaneous equations consists of two (or more) equations whose variables each take the same value at the same time. Provided all the equations are independent, and a solution is possible, then $n$ simultaneous equations containing $n$ variables will have a unique solution. There are several ways of solving simultaneous equations, the most common being by a combination of substitution and elimination.

### Substitution

A substitution in algebra is done by replacing one expression, or part of an expression, by something of equivalent value so that the overall truth of the original expression is unchanged.

**Example**

Given that $4y + 3x = 22$ and $y = 2x$, the 2nd expression can be substituted for $y$ in the 1st expression to give $4(2x) + 3x = 22$, which simplifies to $11x = 22$

### Term

The terms in a simple algebraic expression are the quantities that are linked to each other by means of $+$ or $-$ signs.

**Example**

The expression $5x^2 + 3x - xy + 7$ has four terms.

# Extension worksheet

This worksheet develops the work on quadratic equations and simultaneous equations by introducing the equation of a circle.

The equation of a circle is given generally in the form $(x - a)^2 + (y - b)^2 = r^2$ where $(a, b)$ are the coordinates of the centre of the circle and $r$ is the radius. The circle below has centre at $(0, 0)$ and radius 2. Hence its equation is $x^2 + y^2 = 4$.

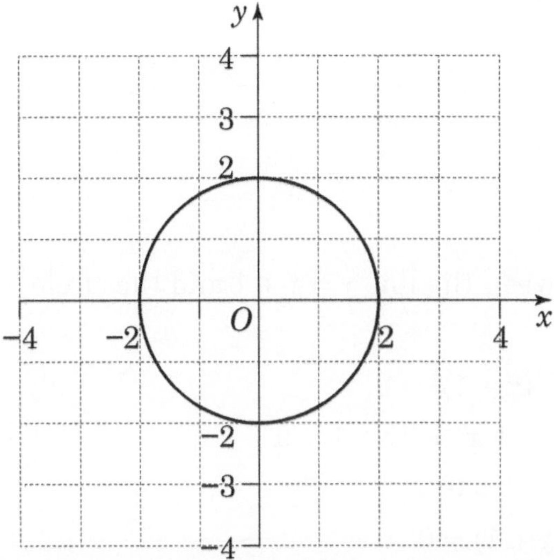

This circle has equation $(x - 2)^2 + (y - 1)^2 = 9$ since its centre is at $(2, 1)$ and it has radius 3.

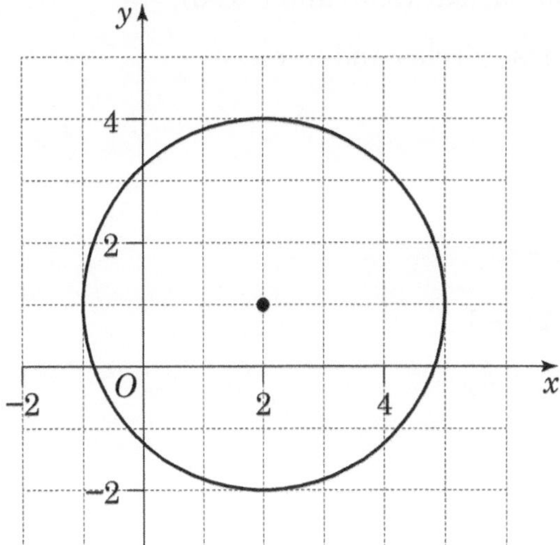

If a straight-line graph is drawn so that it intersects the circle at two points, you can solve to find the two points by *substitution* of the line equation into that of the circle. Effectively, you are solving the two equations simultaneously.

## Example

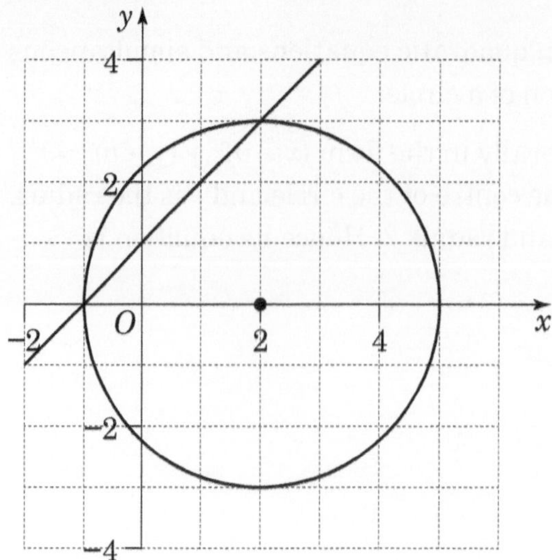

Find the points of intersection between the line $y = x + 1$ and the circle $(x - 2)^2 + y^2 = 9$.

| | |
|---|---|
| Substitute for $y$: | $(x - 2)^2 + (x + 1)^2 = 9$ |
| Expand both brackets: | $x^2 - 4x + 4 + x^2 + 2x + 1 = 9$ |
| Form a quadratic: | $2x^2 - 2x - 4 = 0$ |
| Factorise and solve: | $2(x - 2)(x + 1) = 0$ |
| | $x = 2$ or $x = -1$ |
| | $y = 3$ or $y = 0$ |

Hence the coordinates of intersection are at (2, 3) and (−1, 0).

Find the points of intersection of these circles and lines.

1. $x^2 + y^2 = 4$ and $y = x + 2$

2. $(x - 1)^2 + (y - 1)^2 = 9$ and $y = x - 3$

3. $(x - 1)^2 + (y + 2)^2 = 9$ and $y = 2 - x$

4. $(x - 2)^2 + (y - 3)^2 = 16$ and $y = 2x + 1$

5. $(x + 2)^2 + (y - 4)^2 = 25$ and $y = 3 - 2x$

6. $\left(x - \dfrac{1}{2}\right)^2 + \left(y - \dfrac{3}{2}\right)^2 = 10$ and $x + y = 3$

(Answers on page 125)

This chapter is in three main sections.

The first section is about proportionality. It starts with a discussion about ratio, before moving on to look at direct and inverse proportionality. Three specific examples of this, currency exchange, map scales and area scale factors, are then considered.

The second section is about percentages. Following a quick revision exercise, students consider percentage increases and decreases, profit and loss, simple and compound interest (the second being an example of repeated percentage change) and calculating income tax.

The last section is about rates of change.

It starts with two exercises about time and time intervals.

This leads on to questions related to speed, distance and time, before the chapter ends by considering other rates of change, such as pressure, density and population density.

## Prerequisite knowledge

Before attempting this chapter, students should:

- have a basic understanding of ratios
- have a basic understanding of percentages
- be familiar with metric units.

# Investigations, activities and puzzles

## A trip to the seaside

A family take a trip to the seaside.

On the way, their average speed is 30 km/h.

On the way back, they encounter traffic and their average speed is only 20 km/h.

What is their average speed for the whole journey?

**Notes on this activity**

This is a very well-known puzzle. Most people's first instinct is to suspect that the answer is 25 km/h, but the answer is actually 24 km/h. This is because they spend more time travelling at the slower speed than at the faster speed and what is important is how long they are travelling at that speed rather than how many miles they travel at that speed.

An interesting feature of the puzzle is that the actual distance to the seaside does not affect the answer. When you do the algebra you will see that the distance cancels out. This means that if the students are struggling to solve the problem you can tell them to make up a distance, so that they can use it in their calculations. A sensible distance to use is 60 km.

If you suppose that the distance is 60 km, you get the following:

time = distance ÷ speed

so,

time to get to the seaside
= 60 ÷ 30 = 2 hours

time to get back home
= 60 ÷ 20 = 3 hours

total distance = 2 × 60 = 120 km

total time spent travelling = 5 hours

average speed for the whole journey = 120 ÷ 5 = 24 km/h

However, if you don't substitute a value for the distance to the seaside and instead call it $d$, you get the following:

time to get to the seaside
$= \dfrac{d}{30}$ hours

time to get back home
$= \dfrac{d}{20}$ hours

total distance = $2d$ km

average speed for the whole journey

$$= \frac{\text{total distance}}{\text{total time}} = \frac{2d}{\dfrac{d}{30} + \dfrac{d}{20}} = \frac{2d}{\dfrac{2d + 3d}{60}} = \frac{2d}{\dfrac{5d}{60}} = \frac{2d^{1} \times 60}{5d^{1}} = \frac{120}{5} = 24 \text{ km/h}$$

## Calculator words

On a calculator, work out $9508^2 + 192^2 + 10^2 + 6$. If you turn the calculator upside down and use a little imagination, you can see the word 'HEDGEHOG'.

Find the words given by the clues below.

1. $19 \times 20 \times 14 - 2.66$ (not an honest man)

2. $(84 + 17) \times 5$ (dotty message)

3. $904^2 + 89\,621\,818$ (prickly creature)

4. $(559 \times 6) + (21 \times 55)$ (what a surprise!)

5. $566 \times 711 - 23\,617$ (eat it quickly)

6. $\dfrac{9999 + 319}{8.47 + 2.53}$ (sit up and plead)

7. $\dfrac{2601 \times 6}{4^2 + 1^2}$; $(2 + 3) \times 11 \times 10^2 + 9 - 1$ (two words for the person in charge)

8. $7 \times 10 \times 100 + 2 \times 9 \times 41$ (you can ring it)

9. $\dfrac{27 \times 2000 - 2}{0.63 \div 0.09}$ (not quite a mountain)

10. $(5^2 - 1^2)^4 - 14\,239$ (a name)

11. $48^4 + 102^2 - 4^2$ (pursuits)

12. $615^2 + (7 \times 242)$ (a small laugh)

13. $(130 \times 135) + (23 \times 3 \times 11 \times 23)$ (wobbly)

14. $164 \times 166^2 + 734$ (almost big)

15. $8794^2 + 25 \times 342.28 + 120 \times 25$ (thin skin)

16. $0.08 - (3^2 \div 10^4)$ (ice house)

17. $235^2 - (4 \times 36.5)$ (shiny surface)

18. $3 \times 17 \times (329^2 + 2 \times 173)$ (unable to stand)

19. $230^2 + \dfrac{1}{2} \times 230 + 30$ (these go on feet)

20. $(23 \times 24 \times 25 \times 26) + (3 \times 11 \times 10^3) - 20$ (help)

21. $(22^2 + 29.4) \times 10$; $(3.03^2 - 0.02^2) \times 100^2$ (four words, Goliath)

22. $1.25 \times 0.2^6 + 0.2^2$ (don't cry)

23. $(3^3)^2 + 2^2$ (wriggler)

24. $14 + (5 \times (83^2 + 110))$ (bigger than a duck)

25. $2 \times 3 \times 53 \times 10^4 + 9$ (opposite to hello, almost!)

26. $(177 \times 179 \times 182) + (85 \times 86) - 82$ (two words, good salesman)

27. $(426 \times 474) + (318 \times 487) + 22\,018$ (goes on a hat)

**Notes on this activity**

This is just a fun activity designed to give students practice at accurately typing a calculation into their calculators. Here are the answers.

1. 5317.34 (HE LIES)
2. 505 (SOS)
3. 90439034 (HEDGEHOG)
4. 4509 (GOSH)
5. 378809 (GOBBLE)
6. 938 (BEG)
7. 918, 5508 (BIG BOSS)
8. 7738 (BELL)
9. 7714 (HILL)
10. 317537 (LESLIE)
11. 5318804 (HOBBIES)
12. 379919 (GIGGLE)
13. 35007 (LOOSE)
14. 4519918 (BIGGISH)
15. 77345993 (EGGSHELL)
16. 0.0791 (IGLOO)
17. 55079 (GLOSS)
18. 5537937 (LEGLESS)
19. 53045 (SHOES)
20. 391780 (OBLIGE)
21. 5134, 91805 (HE IS SO BIG)
22. 0.04008 (BOOHOO)
23. 733 (EEL)
24. 35009 (GOOSE)
25. 3180009 (GOODBIE)
26. 5773534 (HE SELLS)
27. 378808 (BOBBLE)

# Glossary

### Average speed

When an object moves through some distance its speed may vary as it travels, but its average speed is found by considering the total time taken and the total distance moved.

The three equations used in working with average speed are:

speed = distance ÷ time

time = distance ÷ speed

distance = speed × time

### Example

A car is driven from Aswan to Cairo, a distance of 780 km, and the journey takes 20 hours. What is the average speed for the journey? The average speed is $780 \div 20 = 39$ km/h.

### Compound interest

The compound interest at the end of a period is calculated using the principal plus any previous interest already earned. This is the usual method for interest calculations in practice.

For example, the total amount owing with compound interest (not just the interest part) is calculated as:

total amount owing =

$$\text{principal} \times \left(1 + \frac{\text{rate of interest (\%)}}{100}\right)^{\text{number of periods}}$$

### Percentage

A value given as a percentage means that the number stated can be represented by a fraction with that number on the top and 100 on the bottom.

% is the symbol for per cent.

### Examples

45 per cent is the fraction $\dfrac{45}{100}$

37% is 37 per cent or $\dfrac{37}{100}$

### Ratio

A ratio is used to compare the sizes of two (or more) quantities.

### Example

Mortar for building a brick wall is made by mixing 2 parts of cement to 7 parts of sand. (The parts may be decided by mass or by volume, as long as the same units are used.)

So, the ratio of cement to sand is 2 to 7, which is also written in the form $2:7$

### Scale

When making a drawing of an object that is meant to be in proportion to the size of the object, and from which measurements can be taken, a scale is used to fix the ratio between the actual measurements on the object and those in the drawing.

### Simple interest

The simple interest at the end of a period is calculated using only the rate of interest on the principal (the original amount). However, this method is not often used in practice.

### Example

If you borrow $300 at 2% per month simple interest, at the end of each month you would owe $6 interest. So, at the end of a year, the total interest owing would be $72 (plus the loan, which still has to be repaid).

total simple interest =

$$\frac{\text{principal} \times \text{rate of interest (\%)} \times \text{number of periods}}{100}$$

### Speed

The speed of a moving object is a measure of the distance travelled by the object in a unit period of time.

Any suitable units may be used for the distance (metres, feet, miles, kilometres, and so on) and the time (hours, minutes, seconds, and so on).

# Extension worksheet

This worksheet develops the work on speed–time graphs and introduces the equations of motion under constant acceleration.

Consider the speed–time graph shown, which plots a car accelerating from 4 m/s to 8 m/s in 15 seconds.

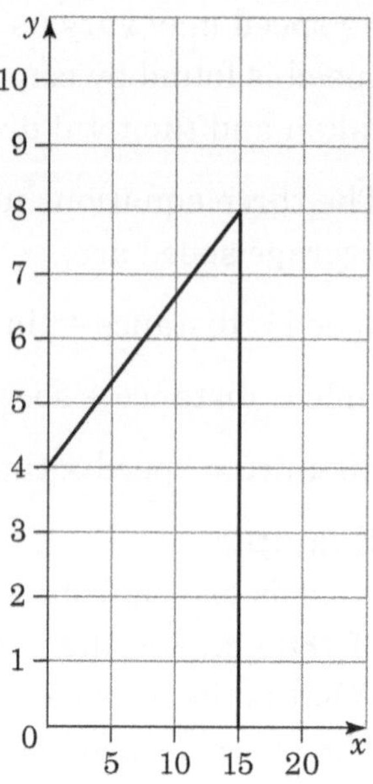

1. Calculate the acceleration.

2. Calculate the distance travelled.

Now, if the *initial speed* is given by $u$ and the *final speed* is given by $v$, a general equation for the acceleration can be found:

gradient of line $= \dfrac{v - u}{t}$ where $t$ is the time taken.

This means that you can write $a = \dfrac{v - u}{t}$.

This equation can be rearranged to give $v = u + at$ which describes the relationship between the four given variables for the general case.

3. Using $u$, $v$, $t$ and $s$ (distance), show that $s = \dfrac{u + v}{2} \times t$.

4. By substituting $v = u + at$ into the equation described in Question **3**, find another equation of the form '$s = ...$', simplifying your answer as much as possible. Which four variables are now connected by your third equation?

5. Rearrange the equation in Question **3** to make $t$ the subject.

6. Substitute your answer to Question **5** into the equation $v = u + at$ and rearrange the resulting equation into the form '$v^2 = ....$'. Which four variables are now connected by this fourth equation?

7. Formulate an equation which connects $s$, $v$, $t$ and $a$ by eliminating $u$ from your system of equations.

Note: The five equations derived here are the equations of motion under constant acceleration. Each one connects *four* of the five variables and hence any unknown can be calculated from *three* other pieces of information.

(Answers on page 125)

# Trigonometry 1

This chapter is all about triangles.

It starts by looking at Pythagoras' theorem, in two and three dimensions.

Students then move on to learn about the three main trigonometric ratios and how to use them to find the length of a side. There is an extra exercise included to cover compound shapes. Students then use inverse ratios to find the size of unknown angles.

The next section shows students how to learn several values of the sin, cos and tan functions, in case a trigonometry question appears on a non-calculator paper. This makes use of Pythagoras' theorem.

The focus then moves on to angles generally.

Students study bearings, followed by angles of elevation and depression. These questions may all involve trigonometry.

The chapter closes by briefly considering scale drawing and three-dimensional trigonometry.

## Prerequisite knowledge

This chapter assumes that students have a basic understanding of:

- square roots
- ratios
- angle facts.

# Investigations, activities and puzzles

## Pythagorean triples

A right-angled triangle can be drawn with side lengths 3, 4 and 5. These are commonly called '3 4 5 triangles'. They are right-angled because $3^2 + 4^2 = 5^2$. We call 3, 4, 5 a Pythagorean triple.

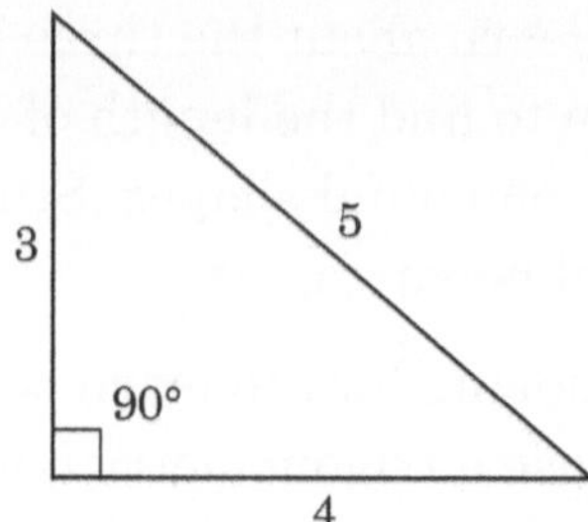

1.  Can you find any other Pythagorean triples?

2.  How many Pythagorean triples (using only whole numbers) exist?

**Notes on this activity**

Students will probably find 6, 8, 10 and 5, 12, 13 fairly quickly.

Discuss the fact that 6, 8, 10 is just an enlargement of 3, 4, 5, but 5, 12, 13 is not.

Clearly, because 6, 8, 10 is an enlargement of 3, 4, 5, this means that there are an infinite number of Pythagorean triples. Are there, however, an infinite number of triples that are not just multiples of each other?

**Extension**

1.  Introduce the idea that Pythagorean triples can be generated by choosing two different whole numbers, $m$ and $n$, and then finding $m^2 + n^2$, $m^2 - n^2$ and $2mn$.

    Here are some questions students could consider:

    - Which values of $m$ and $n$ give 3, 4, 5?
    - Which values of $m$ and $n$ give 5, 12, 13?
    - Use this method to generate other triples. What do you notice?
    - Can you show algebraically why this method works?
    - Do you think that every Pythagorean triple can be generated in this way? Why or why not?

2.  Tell the class about Fermat's last theorem. What can they find out about it?

## Measuring the heights of trees

You can use trigonometry to measure the height of a tree or a tall building.

Stand looking up at the top of the tree and measure the angle of elevation through which you are looking. You can do this using a clinometer.

Then use the tangent ratio to work out the height of the tree.

What happens if the angle you are looking through is 45°?

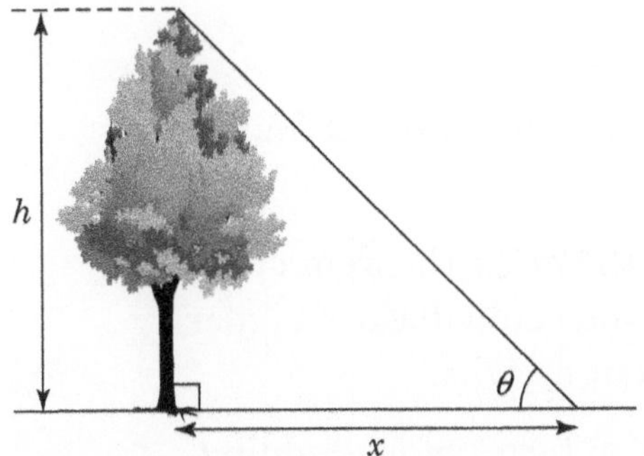


**Notes on this activity**

For example, if you are 10 metres away from the tree and the angle you measure is 50°, the calculation becomes $\tan 50° = \dfrac{h}{10}$, so $h = 10 \tan 50° = 11.92$ m

Then you need to add the height above the floor at which you are holding your angle measuring device. If you do not have a clinometer you can use a square piece of paper and cut along its diagonal. This will give you a 45° angle. If you walk away from the tree, while looking up through a 45° angle until the paper lines up with the top of the tree, the distance you are away from the tree will be the same as the height of the tree.

It is best to do this activity during the summer term or early autumn when the ground is dry!

## Pythagoras in four dimensions

Students will be aware that Pythagoras' theorem can be extended into three dimensions, so that

$$a^2 + b^2 + c^2 = d^2$$

Can the theorem be extended into four dimensions so that

$$a^2 + b^2 + c^2 + d^2 = e^2 \,?$$

If so, what does this mean?

**Notes on this activity**

Four-dimensional shapes are something that students find very interesting.

Although the actual application of Pythagoras' theorem may be too complicated to consider in detail, you could discuss higher dimensions in general as an enrichment activity.

Extra dimensions could be 'explained' in terms of just adding another coordinate.

You may, however, wish to draw their attention to books such as *Flatland* by Edwin A. Abbott, and also consider the 'polytopes' discovered by Alicia Boole Stott in the late 19th century. Alicia Boole Stott is an interesting example of a female mathematician that you could discuss.

# Glossary

### Bearing

The bearing of position B from position A is the direction (usually given as a compass angle) in which someone travelling in a straight line from A to B must go. Care should be taken when using 'from' and 'to' in this work. Bearings are measured clockwise from north.

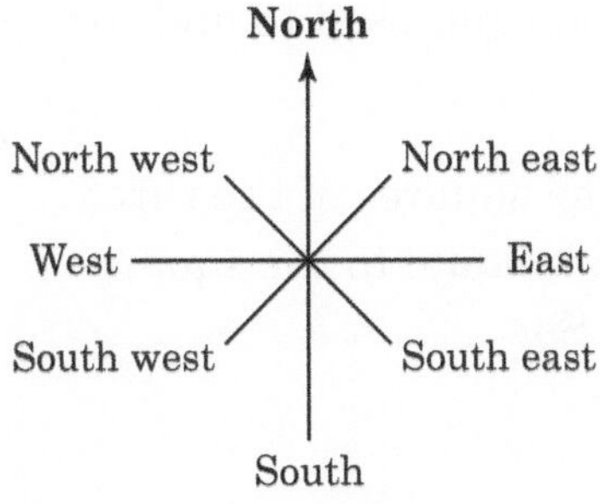


The 8 eight points of the compass by name

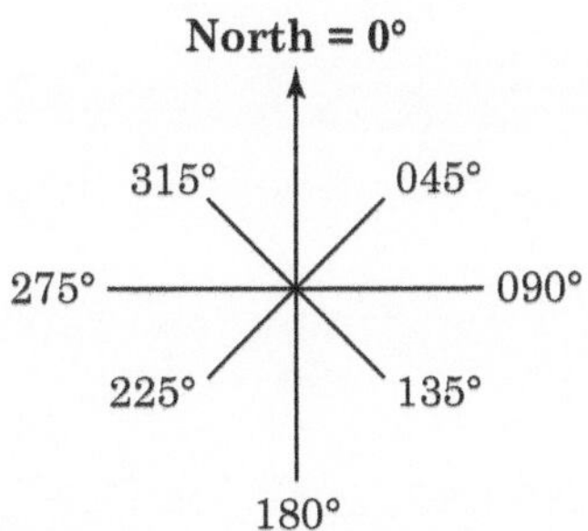


The 8 eight points of the compass measured in degrees clockwise from north

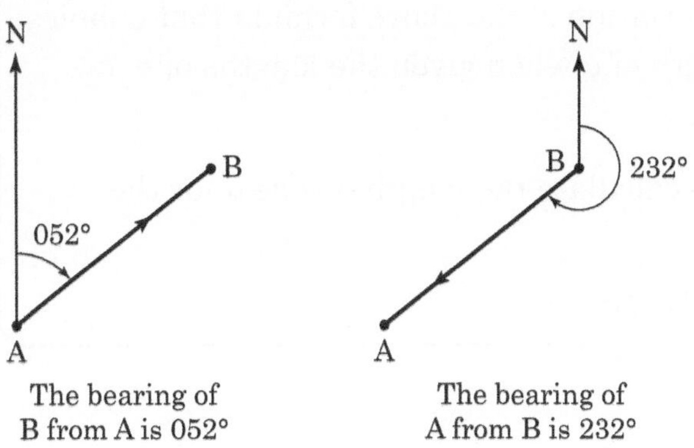


The bearing of B from A is 052°

The bearing of A from B is 232°

### Hypotenuse

The hypotenuse is the edge of a right-angled triangle that is opposite the right angle. It is also the longest edge of that triangle.

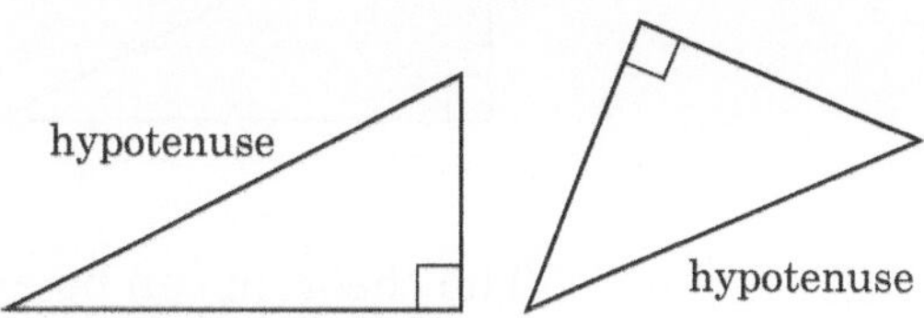


### Triangle

A triangle is a polygon that has three sides. Its three interior angles add up to 180 degrees. Triangles are described with reference to their sides or their vertices (or both).

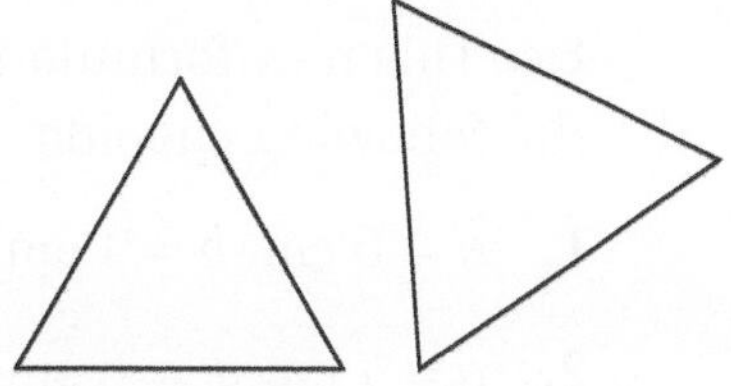

### Trigonometry

Trigonometry is the study of triangles with regard to their measurements and the relationships between those measurements using trigonometric ratios; it also goes on to deal with trigonometric functions.

# Extension worksheet

This worksheet extends the work on Pythagoras' theorem into three dimensions.

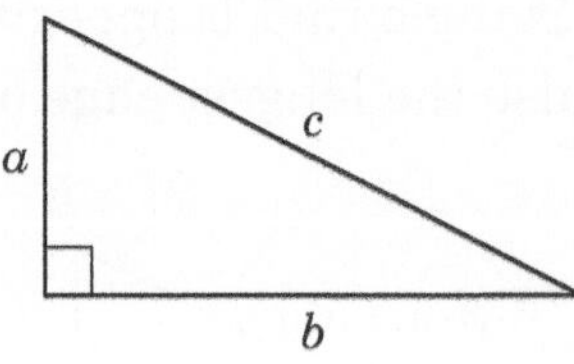

Pythagoras' theorem states that in a right-angled triangle, the sum of the squares of the shorter two sides is equal to the square of the hypotenuse. So

$$a^2 + b^2 = c^2$$

The theorem can be extended to three-dimensions and used to find the length of the long diagonal in a cuboid.

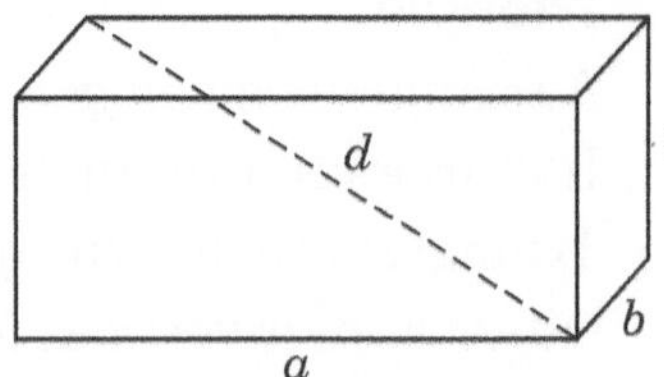

Here, the sum of the squares of the three sides of the cuboid is equal to the square of the long diagonal. So

$$a^2 + b^2 + c^2 = d^2$$

Use this new formula to work out the length of the long diagonal, $d$ for the following cuboids.

1. $a = 6$ cm, $b = 4$ cm and $c = 3$ cm

2. $a = 7$ cm, $b = 3$ cm and $c = 2$ cm

3. $a = 10$ cm, $b = 5$ cm and $c = 6.5$ cm

4. $a = 25$ cm, $b = 12.2$ cm and $c = 11$ cm

5. $a = 8.6$ cm, $b = 4.8$ cm and $c = 2.3$ cm

## Extension

6. Write down a suitable variation of the above formula that enables you to calculate the length of $a$ when given the lengths of $b$, $c$ and $d$.

7. Use this new formula to calculate the length of side $a$ for the cuboid when

   $b = 7$ cm, $c = 6$ cm and $d = 18$ cm.

(Answers on page 125)

This is a chapter about two-dimensional and three-dimensional shapes.

The first section deals with area problems involving triangles, rectangles, parallelograms, trapezia and kites. It is assumed that students already know the basics, so the questions are of a more challenging nature.

Circles are then considered. After a few basic questions, the focus quickly turns to applied, worded questions. Students then learn how to find lengths of arcs and areas of sectors.

The chapter then moves on to three-dimensional shapes:

- prisms
- cylinders
- pyramids
- cones
- spheres

Volume and surface area are both considered.

The chapter then closes with a more general look at volume and surface area, including a section on areas and volumes of similar shapes and objects.

### Prerequisite knowledge

This chapter assumes that students have a basic understanding of:

- areas of basic shapes, such as rectangles and triangles
- basic angle facts
- metric units
- ratios.

# Investigations, activities and puzzles

## Area and perimeter

1.  Consider the following two rectangles.

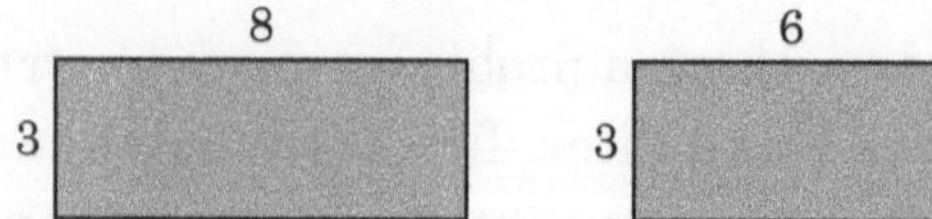

    The first rectangle has a perimeter of 22 units and an area of 24 square units.

    The second rectangle has a perimeter of 18 units and an area of 18 square units.

    How many rectangles can you find whose perimeter and area have the same numerical value?

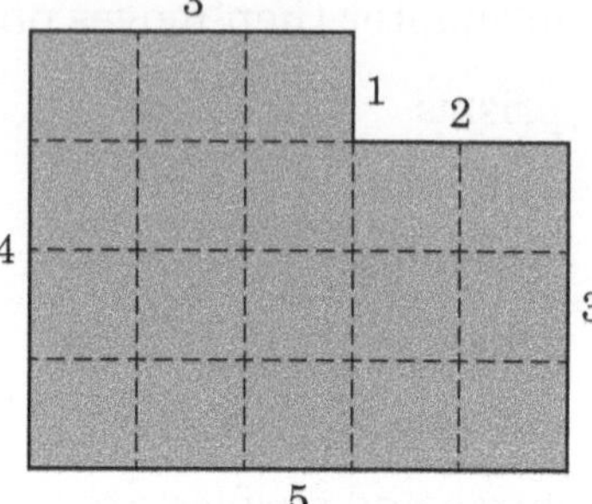

2.  Extend this idea to consider rectangles with pieces removed.

    How does this affect your results?

---

**Notes on this activity**

If a rectangle has a length $a$ and a width $b$, the area is $ab$ and the perimeter is $2(a + b)$.

If the perimeter and area have the same numerical value then

$$2(a + b) = ab$$
$$2a + 2b = ab$$
$$ab - 2a = 2b$$
$$a(b - 2) = 2b$$
$$a = \frac{2b}{b - 2}$$

This means that $a$ will be a whole number when $b - 2$ is a factor of $2b$. This should help students find solutions to the problem.

---

**Extension**

You could extend this activity by considering the following:

*   What happens if you do this with right-angled triangles or equilateral triangles?
*   Could this idea work with a semi-circle?
*   How about three-dimensional shapes? Can you make cuboids, spheres or cylinders whose volumes are numerically equal to their surface area?
*   Can you find any connection between the square with equal area and perimeter and the circle with equal area and perimeter? How about the equilateral triangle with equal area and perimeter?

## Square units

Can any of your students write their name in a square of 0.01 m²?

### Notes on this activity

When you ask students this question, you will probably find that most of them attempt to write their name inside a square of side length 1 cm. This is because they do not realise how the unit conversion rules change when you are dealing with square units.

## Obtuse triangles

For many triangles, the reason why the area is $\frac{1}{2} \times$ base $\times$ height is easy to see, because the triangle can be thought of as being half of a rectangle.

For example, it is easy to see here that the triangle is half of the size of the rectangle in which it is contained.

The relationship for obtuse-angled triangles, however, it is not so obvious.

Is the area of the triangle on the right still half of the rectangle in which it is contained?

If so, how can you show this?

### Notes on this activity

The formula does still work and this can be shown in two neat ways.

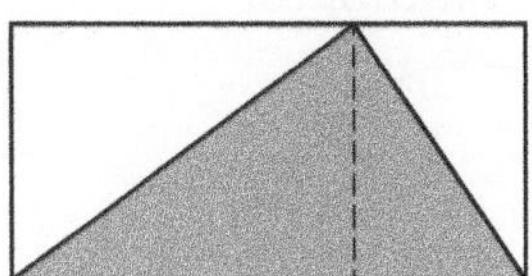

1. Although the obtuse triangle is not obviously half of a rectangle, the following diagram shows that it is half of a parallelogram.

   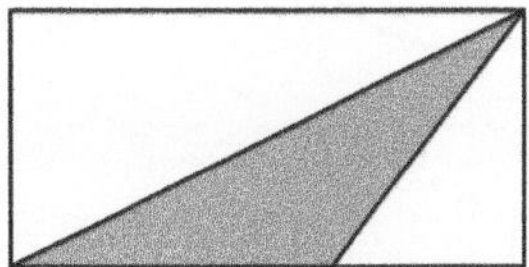

   Since the formula for the area of a parallelogram is the same as the formula for the area of a rectangle, you can see that the formula for the area of a triangle is applicable.

2. The area of the obtuse triangle is

   $$\frac{1}{2}(b+x)h - \frac{1}{2}xh = \frac{1}{2}bh + \frac{1}{2}xh - \frac{1}{2}xh$$

   $$= \frac{1}{2}bh$$

   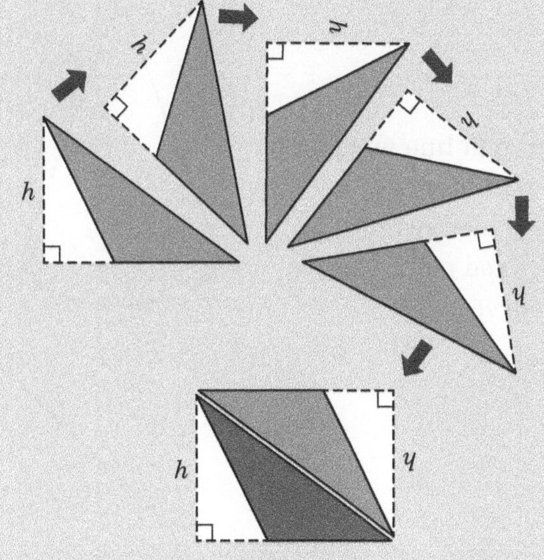

   as required.

# Glossary

## Arc

An arc of a circle is any piece of the curve that makes the circle.

$$\text{length of arc} = \frac{\text{angle (in degrees) of arc at centre}}{360}$$

$$\times \text{ circumference of full circle}$$

## Chord

A chord of a circle is any straight line drawn across a circle, beginning and ending on the curve making the circle. A chord that passes through the centre is also a diameter.

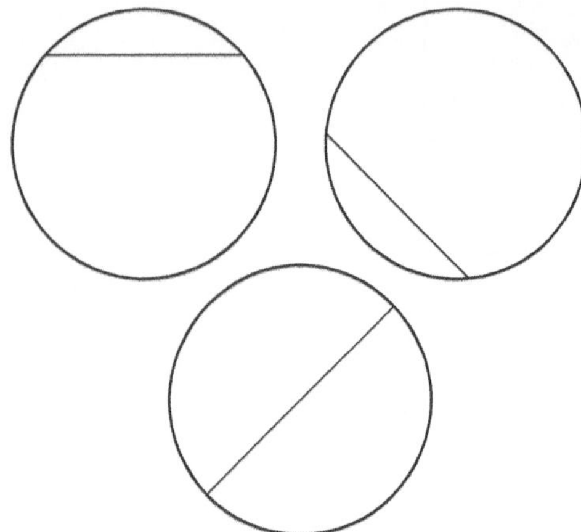

## Circle

A circle is either a closed curve (a line which curves around and joins up with itself) in a plane (a flat surface), which is everywhere the same distance from a single fixed point, or it is the shape enclosed by that curve.

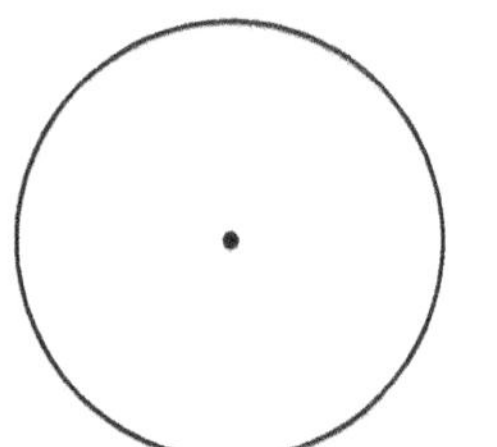

$$\text{area of circle} = \pi \times \text{ radius } \times \text{ radius}$$

$$= \pi r^2$$

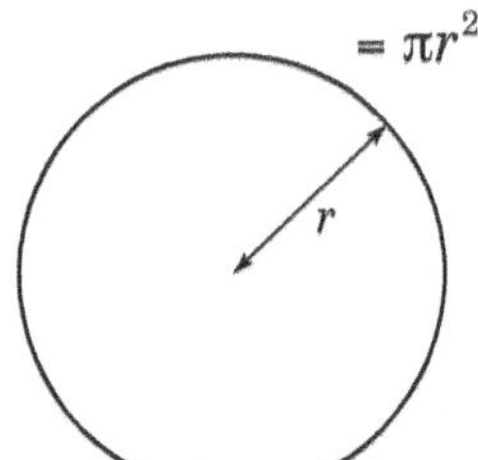
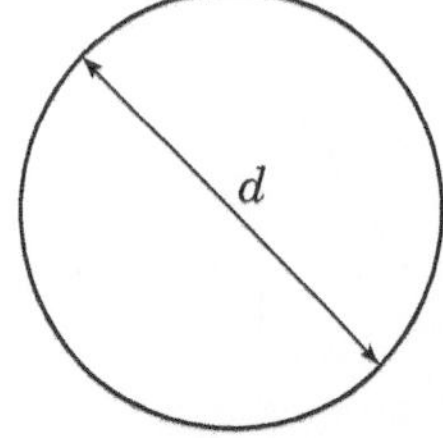

## Circumference

The circumference of a circle is the distance measured around the curve that makes the circle.

$$\text{circumference} = \pi \times \text{diameter} = \pi d$$

## Compound shape

A shape made by joining two or more other shapes together.

## Cone

A cone is the three-dimensional shape formed by a straight line when one end is moved around a simple closed curve while the other end of the line is kept fixed at a point that is not in the plane of the curve.

$$\text{volume} = \frac{1}{3}\pi r^2 h$$

where $h$ is the perpendicular height

$$\text{curved surface area} = \pi r l$$

where $l$ is the slant height

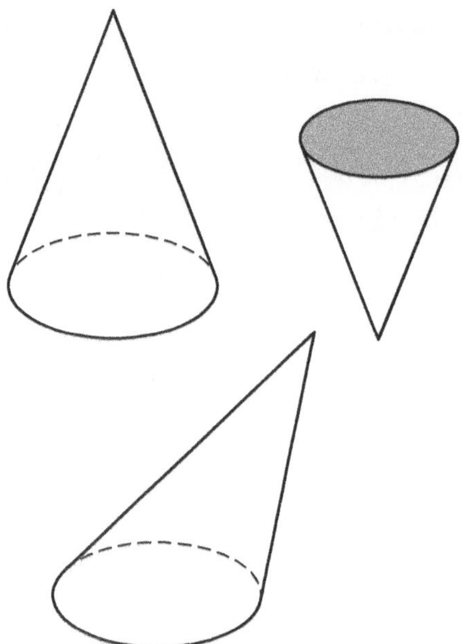

## Cuboid

A cuboid is a prism with six faces that are all rectangles.

$$\text{volume} = l \times b \times h$$

$$\text{surface area} = 2 \times (bl + hl + bh)$$

## Cylinder

A cylinder can be thought of a prism with a circular cross-section.

volume $= \pi r^2 h$

## Diameter

A diameter of a circle is a straight line passing through the centre that touches the curve forming the circle at each of its ends.

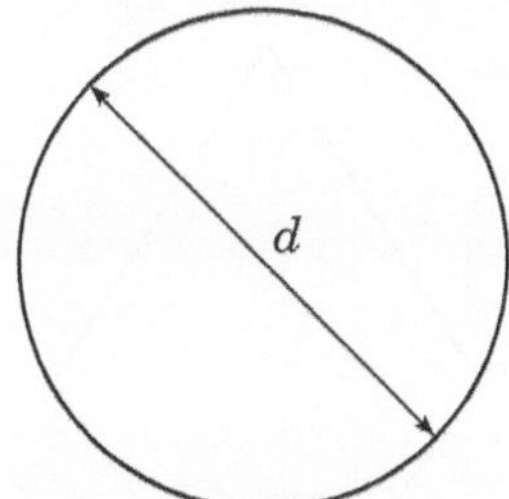

## Hemisphere

A hemisphere is one half of a sphere.

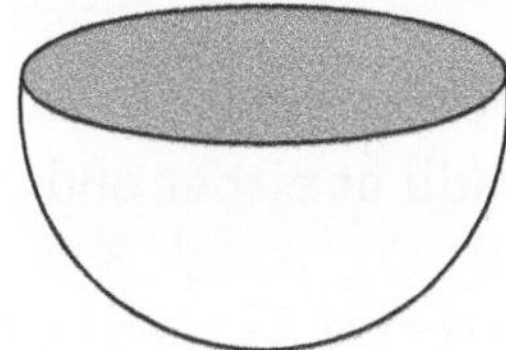

## Kite

A kite is a quadrilateral that has two pairs of adjacent sides (i.e. sides that are next to each other) of the same length, and no interior angle larger than 180°. It has one line of symmetry and its diagonals cross each other at right angles.

## Parallelogram

A parallelogram is a quadrilateral that has two pairs of parallel sides. It has rotational symmetry of order 2, and its diagonals bisect each other. Usually one pair of sides is longer than the other pair, no interior angle is a right angle and it has no lines of symmetry.

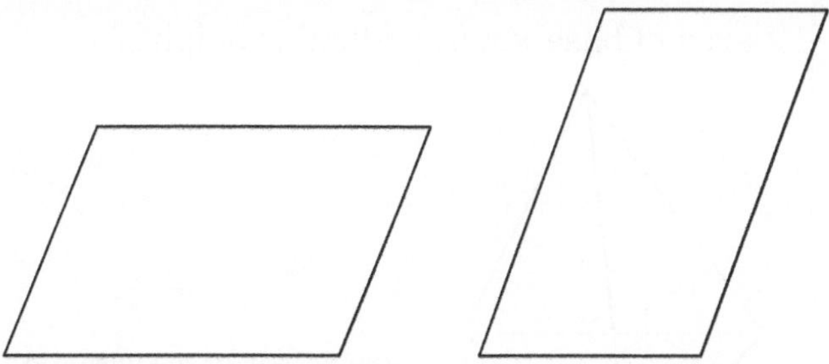

## Polygon

A polygon is a shape completely enclosed by three or more straight sides that do not cross one another. The word is not often used for shapes having fewer than 5 edges. Polygons are named by the number of edges or angles they have.

### Example

The polygon below is a regular pentagon. It has five sides of equal length.

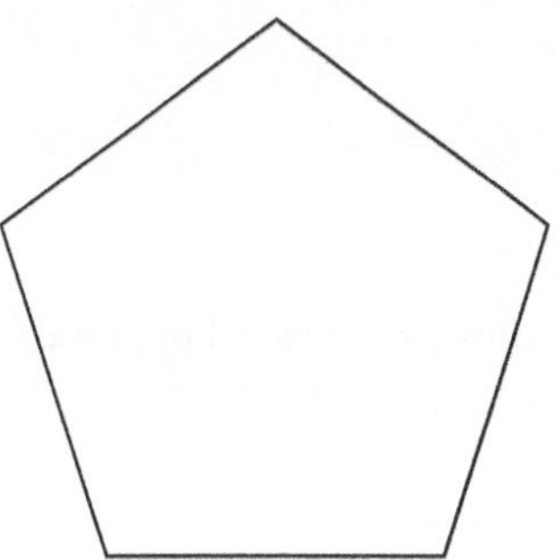

## Prism

A prism is a three-dimensional shape with uniform cross-section.

volume $= A \times l$

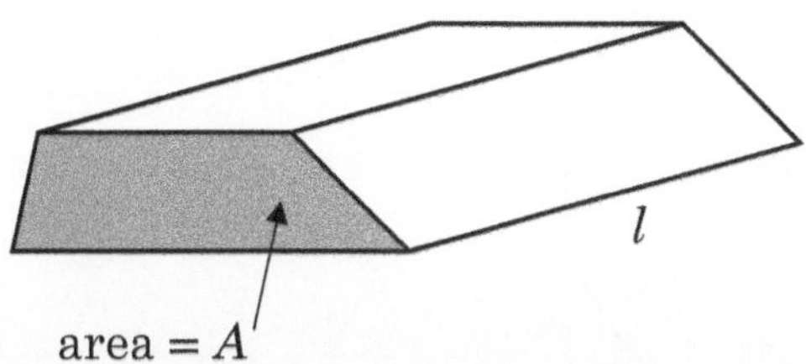

## Pyramid

A pyramid is a polyhedron in which one face is a polygon and all the other faces are triangles meeting at a common vertex. The pyramid is named after the polygon that forms the face from which the triangles start.

volume of a pyramid =
$\frac{1}{3}$ × area of base × perpendicular height

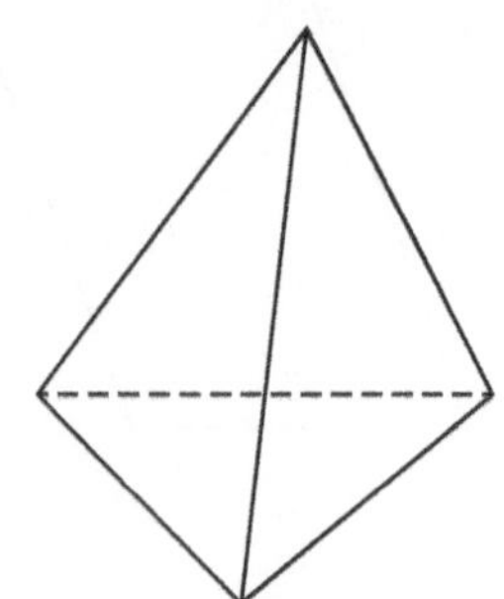

triangular-based pyramid
(tetrahedron)

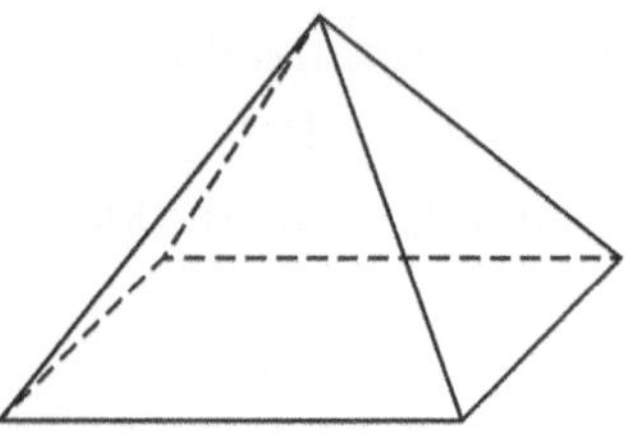

square-based pyramid

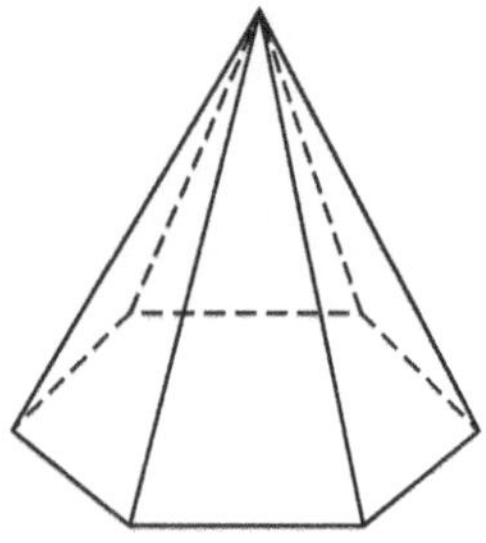

hexagonal-based pyramid

## Radius

The radius of a circle is the distance from the centre to the curve that makes the circle.
A radius is any straight line from the centre to the curve.

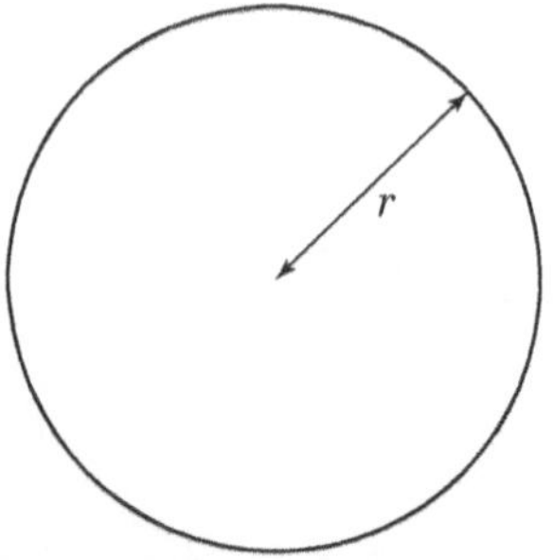


## Rectangle

A rectangle is a quadrilateral in which every interior angle is a right angle.

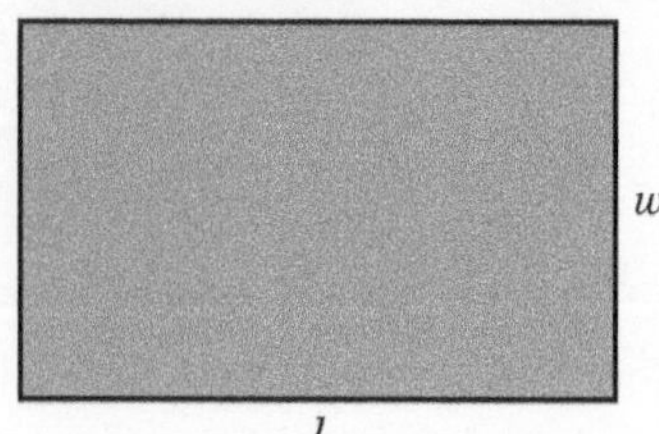


area = $l \times w$

where $l$ and $w$ are the length and width, respectively.

## Rhombus

A rhombus is a quadrilateral whose sides are all the same length. Its diagonals bisect each other at right angles. The diagonals are also lines of symmetry.

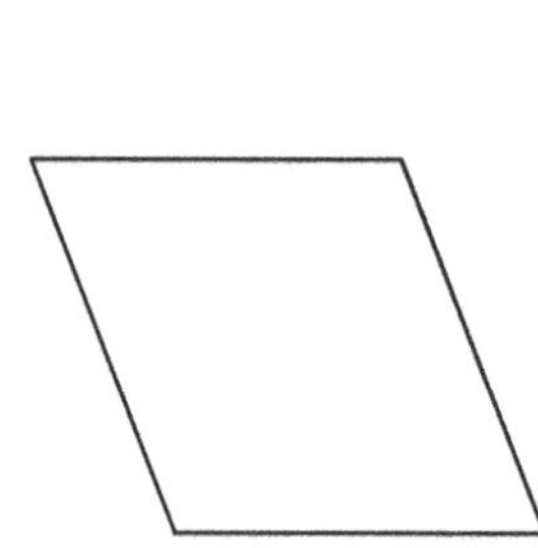

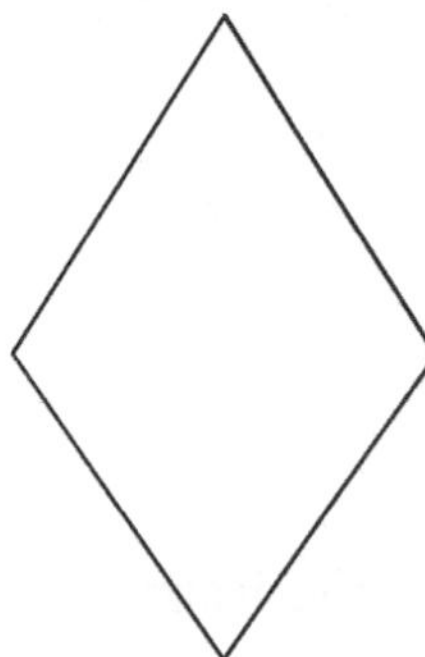

## Sector

A sector of a circle is the shape enclosed between an arc and the two radii at either end of that arc.

area of sector = $\dfrac{\text{angle (in degrees) of sector at centre}}{360}$
× area of full circle

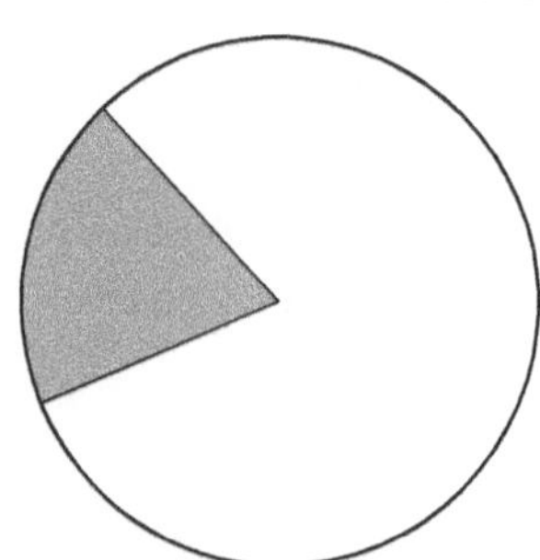

## Similar

Geometrical figures are said to be similar if they are the same in shape but different in size. One shape is an enlargement of another. Corresponding angles in each shape will be the same size. All of these triangles are similar:

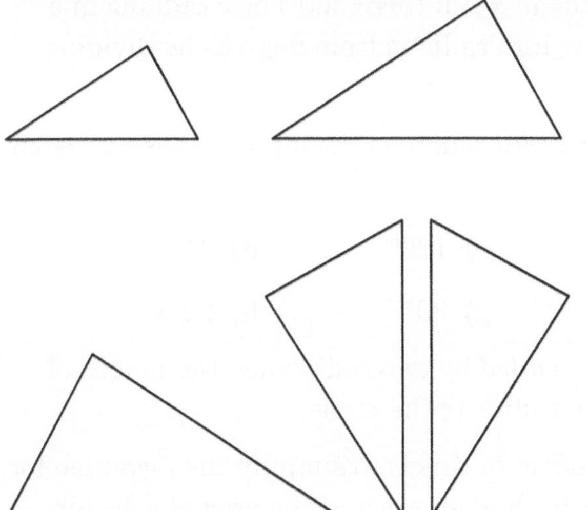

## Sphere

A sphere is either the shape of a surface in three dimensions that is everywhere the same distance from a single fixed point, or it is the solid shape enclosed by that surface. The balls used to play most games are spheres.

$$\text{volume} = \frac{4}{3}\pi r^3$$

$$\text{surface area} = 4\pi r^2$$

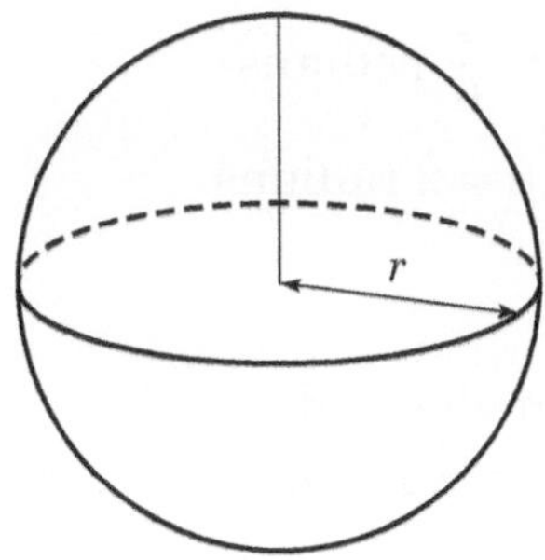

## Trapezium

A trapezium is a quadrilateral with only one pair of parallel sides.

$$\text{area} = \frac{1}{2}(a + b) \times h$$

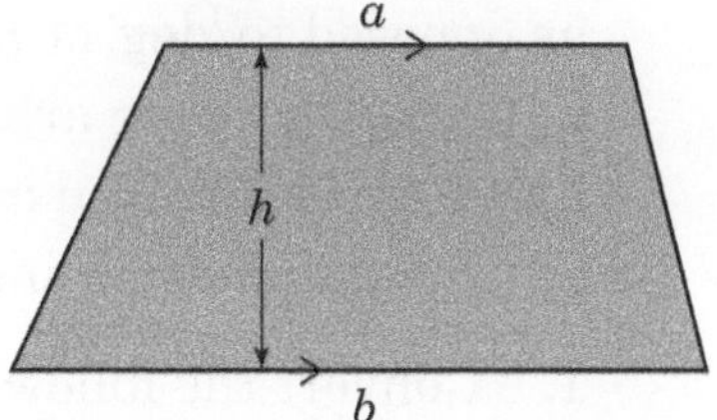

## Extension worksheet

This worksheet develops the work on sector area and arc length by introducing radian measure for angles.

An alternative measure for angle is the *radian* ('rad' or *r* on a calculator as opposed to 'deg' or *d*). One radian is approximately 57° but, more importantly, there are $2\pi$ radians in a full turn (360°) or $\pi$ radians in a half turn (180°). You can convert into radians from degrees by dividing by 180 and multiplying by $\pi$.

1. Convert the following angles into radians (leaving your answer as an exact multiple of $\pi$):

   **a)** 90°        **b)** 60°        **c)** 120°        **d)** 45°

   **e)** 270°       **f)** 135°       **g)** 30°         **h)** 22.5°

   In a circle, one radian is *subtended* by two radii when the length of the minor arc is equal to the radius of the circle.

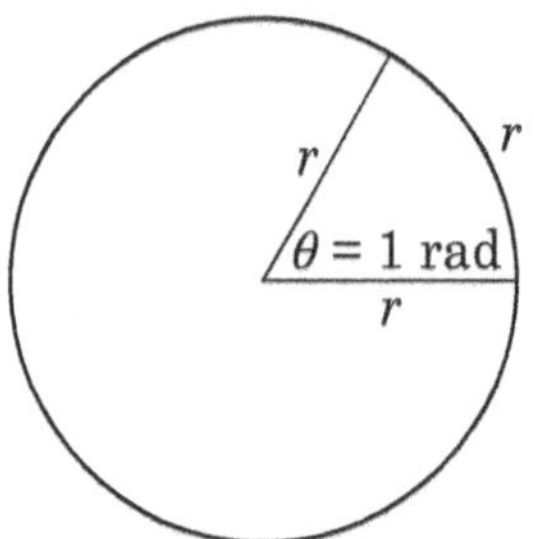


The effect of this is to simplify the formulae for the length of an arc and the area of a sector.

When working in radians (angle $\theta$), the formula for the length of an arc becomes

$$l = r\theta$$

and the formula for the area of a sector becomes

$$A = \frac{1}{2}r^2\theta$$

2. Work out **(i)** the length of the minor arc, and **(ii)** the area of the sector in the following cases.

   **a)** $r = 4$ cm, $\theta = \dfrac{\pi}{2}$ radians        **b)** $r = 6$ cm, $\theta = \dfrac{\pi}{6}$ radians

   **c)** $r = 10.5$ cm, $\theta = \dfrac{\pi}{3}$ radians      **d)** $r = 3.2$ cm, $\theta = 2$ radians

   **e)** $r = 1.63$ cm, $\theta = 3.5$ radians

3. Work out the length of the radius for the given angles and arc lengths (round answers to three significant figures, where appropriate).

   **a)** $l = 6$ cm, $\theta = 3$ radians        **b)** $l = 8$ cm, $\theta = 2$ radians

   **c)** $l = 5$ cm, $\theta = \dfrac{\pi}{4}$ radians        **d)** $l = 4.5$ cm, $\theta = \dfrac{2\pi}{3}$ radians

   **e)** $l = 7.8$ cm, $\theta = \dfrac{\pi}{12}$ radians

4. Work out, correct to three significant figures, the area of the sector formed in each of the parts of Question **3**.

(Answers on page 126)

The attention now turns back to algebra for a short chapter that is mostly concerned with quadratic equations.

The chapter opens with a section on factorising, which includes:

- simple quadratic expressions
- difference of two squares
- simple cubic expressions.

Students are then shown how to solve quadratic equations by:

- factorising
- using the quadratic formula
- completing the square.

Then there is an exercise on problems that can be solved using quadratic equations.

The chapter closes by looking at nonlinear simultaneous equations.

## Prerequisite knowledge

This chapter assumes that students can already:

- use algebra confidently
- solve pairs of linear simultaneous equations
- factorise linear expressions.

## Investigations, activities and puzzles

### Maximum box

You can make a box (without a lid) by cutting squares from the corners of a big square piece of card and folding up the sides.

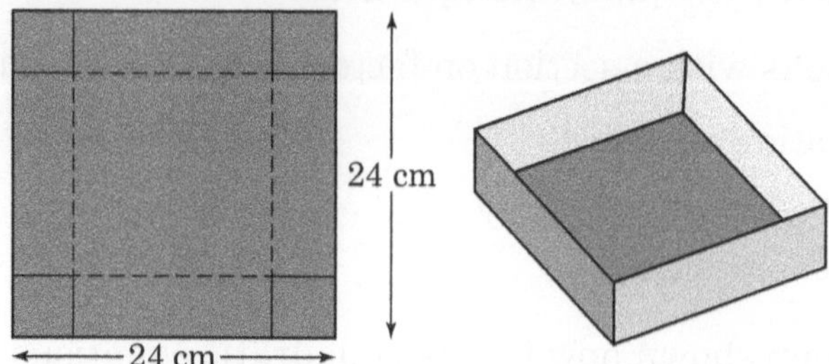


If the sides of the square are 24 cm long, what size corners should you cut out so that the volume of the box is as large as possible?

Try different sizes for the corners and record the results in a table.

| Length of the edges of the corner square (cm) | Dimensions of the open box (cm) | Volume of the box (cm³) |
|---|---|---|
| 1 | 22 × 22 × 1 | 484 |
| 2 | | |
| | | |
| | | |

Now consider boxes made from different-sized cards: 15 cm by 15 cm and 20 cm by 20 cm.

What size corners should you cut out this time so that the volume of the box is as large as possible? Is there a connection between the size of the corners cut out and the size of the square card?

**Notes on this activity**

This is a very popular investigation related to volume, which is covered in this chapter. It is possible to solve this problem using calculus, so you might want to revisit it when teaching Chapter 9.

**Extension**

Consider the same investigation but with card that is not a square.

For example, start with a piece of rectangular card where the length is twice the width. Again, for the maximum volume, is there a connection between the size of the corners cut out and the size of the original card?

## Maximum cylinder

A rectangular piece of paper has a fixed area of 60 cm².

It could, for example, be 5 cm × 12 cm.

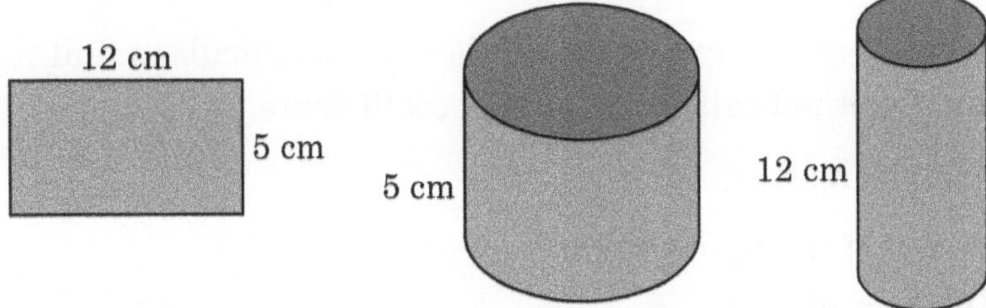


This paper can make a hollow cylinder of height 5 cm or of height 12 cm.

Work out the volume of each cylinder. Which is greater?

What happens if you keep the area of the rectangle at 60 cm² but change the dimensions?

How can you obtain the cylinder with the greatest volume?

### Notes on this activity

This is more complicated than the maximum box example, so it might not be appropriate for all advanced students.

What would happen if you changed the size of the fixed area?

Can you generalise the problem?

### Extension

Rather than fixing the area of the piece of paper, you could instead fix the perimeter.

How would this change the problem?

## Glossary

### Difference of two squares

Any algebraic expression of the form $a^2 - b^2$ can be factorised into $(a + b)(a - b)$.

**Example**

$x^2 - 9$ is $x^2 - 3^2$ which is $(x + 3)(x - 3)$

### Quadratic equation

A quadratic equation is an equation involving an expression, or expressions, containing a single variable, of degree 2.

**Examples**

$x^2 + 3x - 5 = 0 \quad 3(x + 1)^2 = 0 \quad 4x^2 - 3x + 4 = 0$

# Extension worksheet

This worksheet develops the work on factorising a quadratic expression into two brackets, by considering how to factorise a cubic expression, such as $x^3 + 4x^2 - 11x - 30$, into three brackets.

If you know one of the brackets in advance, it is not too difficult to find the other two using a method called comparing coefficients.

This works in the following way.

Say you know that one of the brackets in the factorisation of $x^3 + 4x^2 - 11x - 30$ is $(x - 3)$.

This means that $x^3 + 4x^2 - 11x - 30 = (x - 3)(ax^2 + bx + c)$ for some values of $a$, $b$ and $c$.

Now you need to find the values of $a$, $b$ and $c$.

Expanding $(x - 3)(ax^2 + bx + c)$ gives $ax^3 + bx^2 + cx - 3ax^2 - 3bx - 3c$.

Collecting like terms gives $ax^3 + (b - 3a)x^2 + (c - 3b)x - 3c$.

But $ax^3 + (b - 3a)x^2 + (c - 3b)x - 3c = x^3 + 4x^2 - 11x - 30$.

You can see immediately then that $a = 1$ and $c = 10$.

Considering the $x$ term gives $c - 3b = -11$, and we know $c = 10$.

Therefore $10 - 3b = -11$ and so $3b = 21$ and $b = 7$.

This means that $x^3 + 4x^2 - 11x - 30 = (x - 3)(x^2 + 7x + 10)$.

You now need to factorise the quadratic in the bracket to completely factorise the cubic.

This gives $x^3 + 4x^2 - 11x - 30 = (x - 3)(x + 2)(x + 5)$.

Factorise the following cubic expressions using the method of comparing coefficients.

1.  $x^3 + 6x^2 - x - 30$ if one of the brackets is $(x + 3)$

2.  $x^3 - 3x^2 - 6x + 8$ if one of the brackets is $(x - 4)$

3.  $x^3 + 5x^2 - 17x - 21$ if one of the brackets is $(x + 1)$

4.  $x^3 - 3x^2 - 18x + 40$ if one of the brackets is $(x - 5)$

5.  $x^3 + 5x^2 - 12x - 36$ if one of the brackets is $(x + 2)$

6.  $x^3 + 9x^2 + 14x - 24$ if one of the brackets is $(x + 6)$

7.  $x^3 + x^2 - 46x + 80$ if one of the brackets is $(x - 2)$

8.  $x^3 + 2x^2 - 11x - 12$ if one of the brackets is $(x + 4)$

Investigate how you might factorise a cubic expression if you are not given any of the brackets.

(Answers on page 126)

The focus of this chapter is circle theorems, but some preparation is required before tackling them.

The chapter opens with a revision of angle properties, including angles in polygons and angles in parallel lines.

Then there is a brief section on symmetry, which looks at:

- reflection and rotational symmetry
- axes of symmetry
- planes of symmetry.

Circle theorems are then covered. These are:

- angle at the centre is twice the angle at the circumference
- angles in the same segment are equal
- opposite angles of a cyclic quadrilateral sum to 180°
- angle in a semi-circle is 90°
- angle between tangent and radius is 90°
- tangents from an external point are equal in length
- the alternate segment theorem
- equal chords are equidistant from the centre
- the perpendicular bisector of a chord passes through the centre.

The chapter closes with a brief section on constructions and nets.

## Prerequisite knowledge

Students should already:

- know basic angle facts
- be able to use a pair of compasses.

# Investigations, activities and puzzles

## Constructing shapes

Construct an equilateral triangle of side length 7 cm using only a pencil, a straight edge, and a pair of compasses.

1.  Draw a line segment AB of length 7 cm.

2.  Set your compasses to 7 cm, and then draw an arc from point A.

3.  Keeping your compasses at 7 cm, draw another arc from B that intersects the first arc.

4.  Let C be the point where the two arcs intersect. Join AC and BC to complete the triangle.

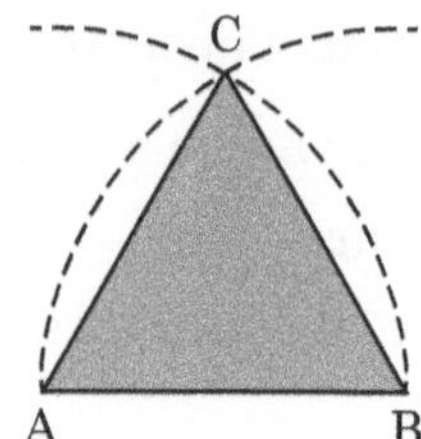

What other shapes can you construct?

### Notes on this activity

This is a great activity. As well as making students more confident using compasses, the idea of being able to accurately construct quite complex shapes such as pentagons and decagons is something that a lot of students really enjoy.

You can use the internet to find out which shapes can be constructed using only a pencil, a straight edge, and a pair of compasses.

There are some very good tutorial videos demonstrating the entire construction process.

### Extension

Since this is an exercise in geometry, there are many ways in which it can be extended.

*   What shapes can be constructed accurately and which ones cannot?
*   Instructions on how to construct good approximations of the ones that can't (for example, heptagon, nonagon) can be found online.

## Nets

Both of these nets can be folded to make a cube.

How many different nets does a cube have?

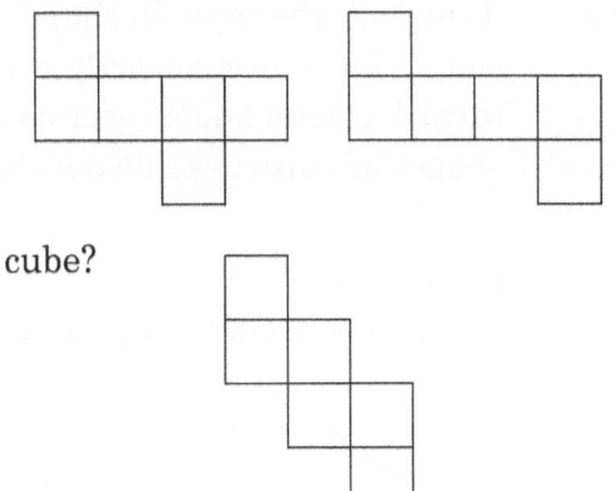

Is this a net for a cube?

> **Notes on this activity**
>
> Explain to students that a rotated version of one net does not count
> as a new net.

## Constructing angles

Using only a pair of compasses, a pencil, and a straight edge, it is
possible to bisect a line and therefore construct a right angle.

It is also possible to accurately bisect an angle, so we can bisect the right
angle as many times as is practical to construct angles of 45°, 22.5°,
11.25° and so on.

You can also construct an equilateral triangle, which means that you can
construct an angle of 60°. Bisecting this gives angles of 30°, 15°, 7.5°
and so on.

What other angles can you construct?

> **Notes on this activity**
>
> This is closely related to the constructing shapes activity.
>
> It has been known for a long time that it is impossible in geometry to
> trisect an angle. However, it is possible to do this using origami. You
> could investigate how to do this with your students.

# Glossary

## Angle

An angle is made when two straight lines cross or meet each other at a point, and its size is measured by how much one line has been rotated away from the other line.

### Examples

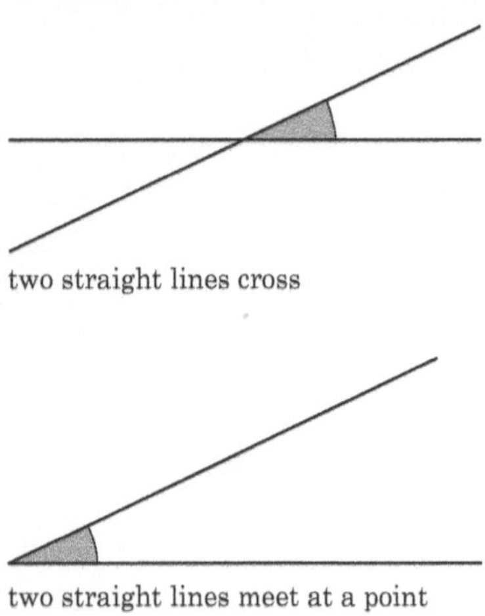

two straight lines cross

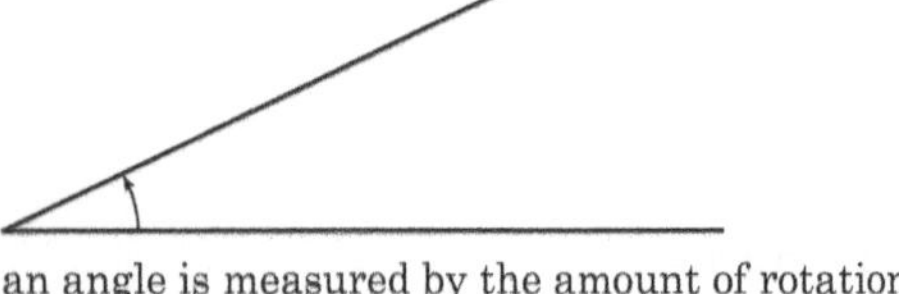

two straight lines meet at a point

an angle is measured by the amount of rotation

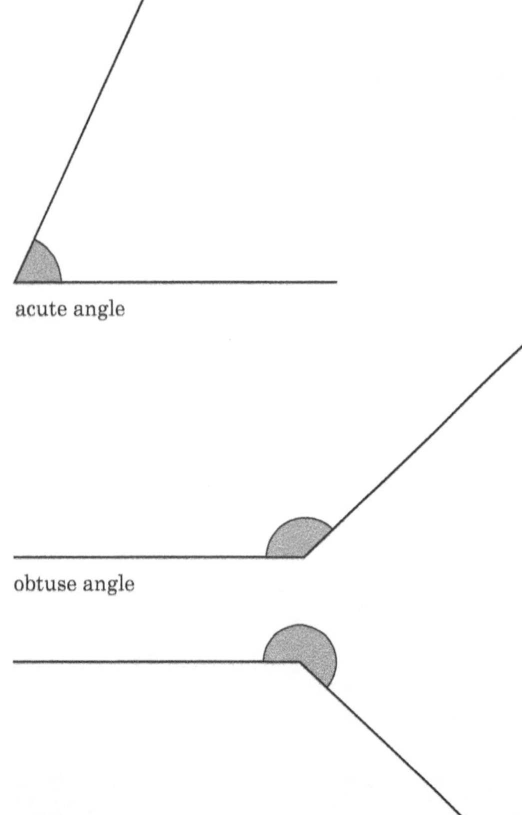

acute angle

obtuse angle

reflex angle

## Congruent

Geometrical figures are said to be congruent if they are the same in shape and size. One shape can be fitted exactly over the other, being turned around and/or over as necessary. Such shapes are also described as being identical or equal.

### Examples

These four triangles are all congruent:

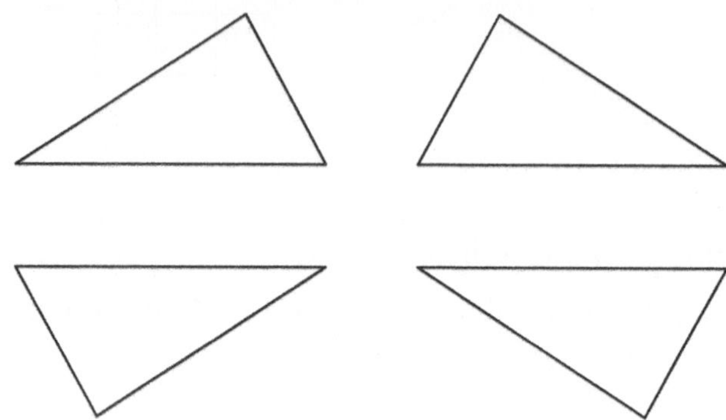

## Line symmetry

Line symmetry is the symmetry of a plane shape (a flat or 2D shape) that can be folded along a line so that one half of the shape fits exactly on the other half. A shape can have several lines of symmetry.

### Examples

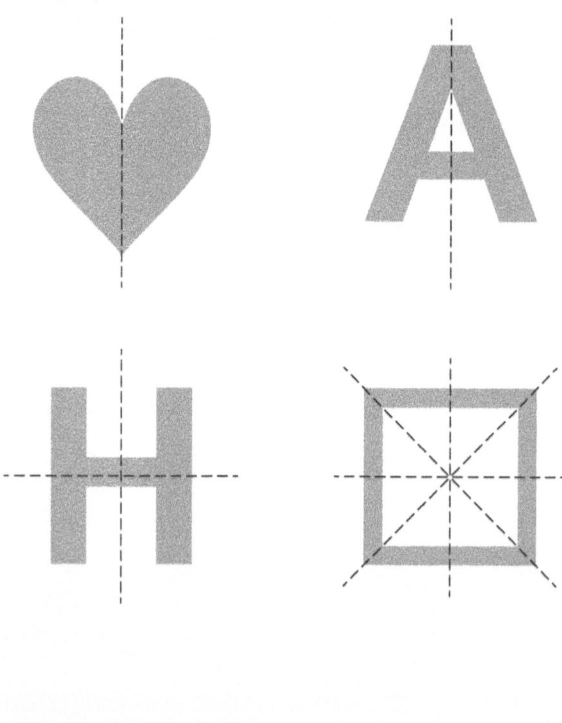

## Net

A net is an arrangement of polygons connected at their sides, all lying in one plane (flat surface), that can be folded up to make a polyhedron. There is always more than one way of doing this.

### Examples

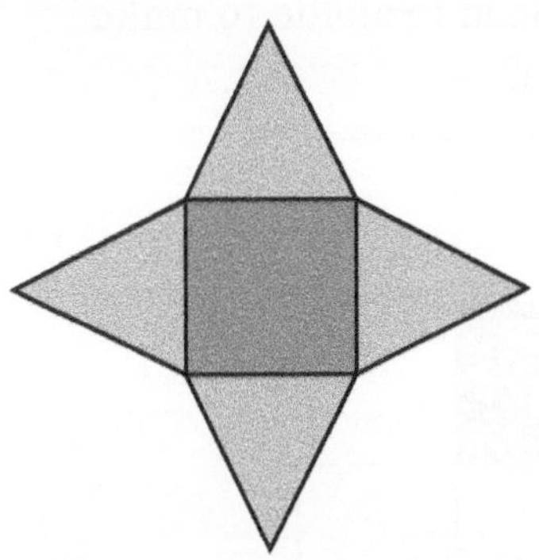

net for a square based pyramid

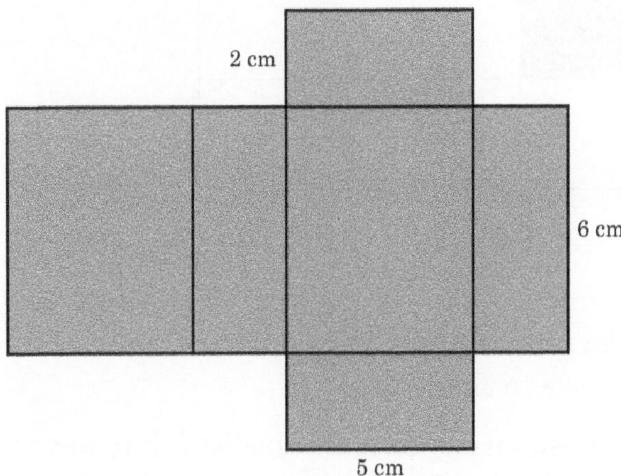



net for a cuboid

## Rotational symmetry

A shape has rotational symmetry if it can be rotated and fitted onto itself somewhere other than in its original position.

### Examples

The centres about which each shape can be rotated is shown by dots.

This shape may be turned about O into three identical positions in one complete rotation. It has rotational symmetry of order 3.

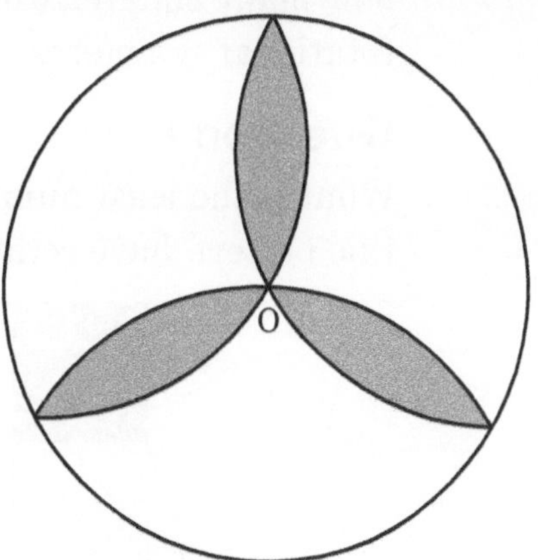

## Symmetry

The word 'symmetry' applied to any object (or situation) means that parts of the object correspond to (or match) other parts in some way.

## Extension worksheet

This worksheet develops the work on rotational symmetry by considering how many squares would need to be shaded to give various orders of rotational symmetry.

### Question 1

What is the least number of extra squares you need to shade to make this pattern have rotational symmetry of order 2?

What about if you want to make it have rotational symmetry of order 4? How many squares would you have to shade then?

What is the maximum number of squares you could shade for it to have rotational symmetry of order 2 and not order 4?

### Question 2

Consider a smaller 3 by 3 square. How many ways are there of shading in squares so that it has rotational symmetry of order 2 but not order 4?

Use the grids on the next page to work this out. Two examples have been done for you.

Can you predict how many ways you can shade a 4 by 4 square so that it has rotational symmetry of order 2 but not order 4?

Is there a connection?

You can use the grids on p44 to help you with this.

What would happen for increasingly bigger grids?

(Answers on page 126)

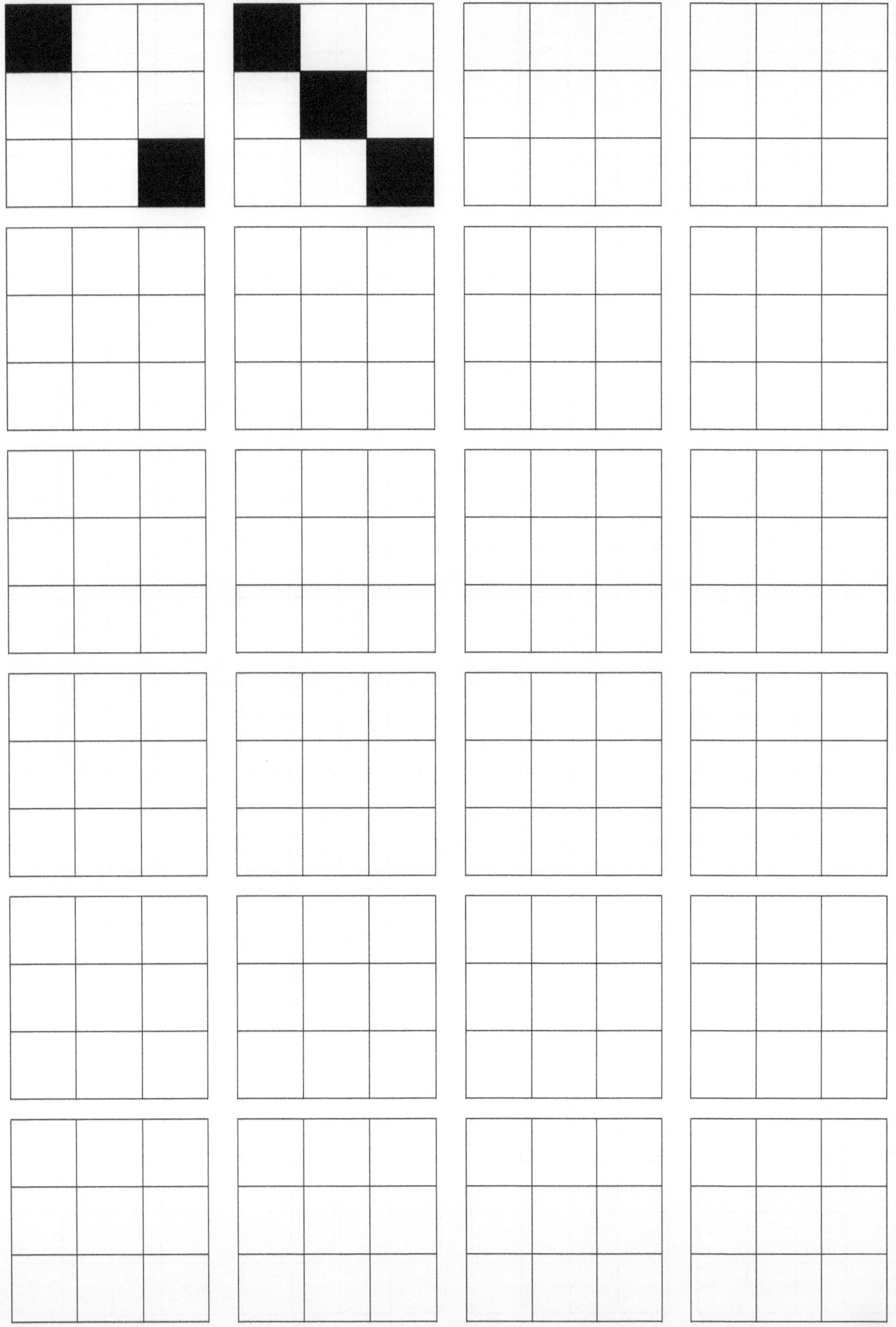

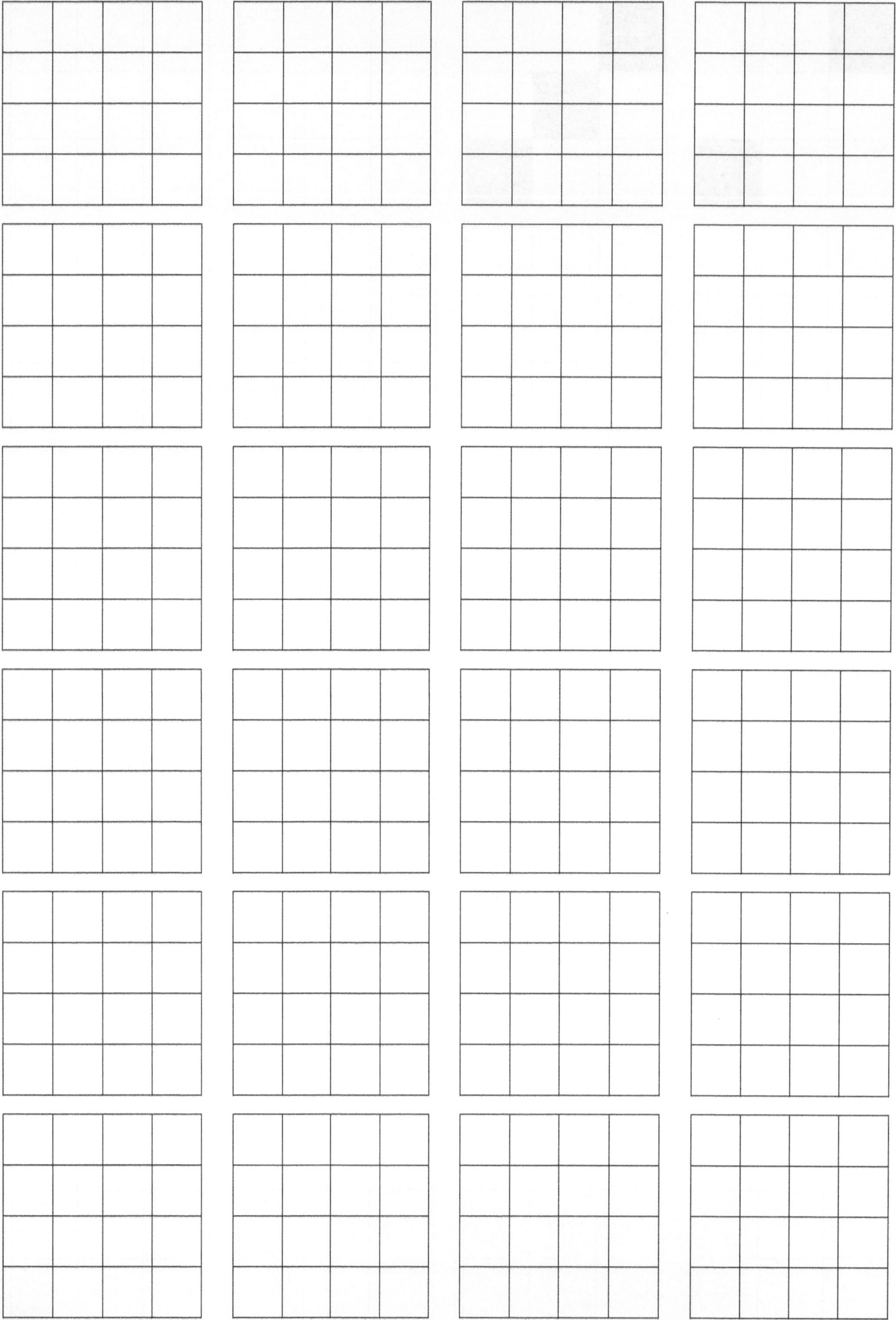

This chapter consists of five main sections.

The first section is about algebraic fractions. Students practise simplifying them, then adding, subtracting, multiplying, and dividing them. The level of challenge here is quite significant. Students then complete an exercise involving solving quadratic equations that contain algebraic fractions, where the questions are presented in the context of real-world situations.

The second section is about changing the subject of a formula. Many of these questions are, again, quite challenging, involving fractions and square roots. In many situations, the letter that is to be the subject of the rearranged formula appears more than once in the original formula.

The third section is on direct and inverse proportion. Algebraically, this section is easier than the previous two, but the emphasis here is on correctly setting out solutions.

After that, there are three exercises on indices. These exercises are again challenging and thorough, involving fractional and negative indices.

The final section concerns two- and three-part inequalities.

## Prerequisite knowledge

Students should already have a good grasp of:

- basic algebra
- basic indices
- basic fractions.

# Investigations, activities and puzzles

## Continued fractions

Look at this continued fraction.

$$\frac{47}{17} = a + \cfrac{1}{b + \cfrac{1}{c + \cfrac{1}{d}}}$$

What are the values of $a$, $b$, $c$ and $d$?

### Notes on this activity

The purpose of this activity is to get the students to practise working with reciprocals. The trick at each stage is to take out the whole number part, flip the fraction part, then start the process again. The finite sequence $a$, $b$, $c$, $d$, ... then becomes a way of encoding an improper fraction.

### Extension

Can you work backwards from the following continued fraction to recreate the original fraction?

$$3 + \cfrac{1}{7 + \cfrac{1}{16}}$$

Recreating an improper fraction from its continued fraction sequence (in this case 3, 7, 16) is an interesting activity in itself. Some well-known irrational numbers have interesting (and sometimes very neat) representations as continued fractions. For example:

$$\sqrt{2} = 1, 2, 2, 2, 2, 2, ...$$

The Golden Ratio $\phi = 1, 1, 1, 1, 1, 1, ...$

$$e = 2, 1, 2, 1, 1, 4, 1, 1, 6, ...$$

However, $\pi$ is more random.

$$\pi = 3, 7, 15, 1, 292, 1, 1, 1, 2, 1, 3, 1, ...$$

## Inverse square laws

There are many inverse square laws in science, such as Newton's law of gravitation.

$$F = G\frac{mM}{r^2}$$

Can you find any other well-known inverse square laws?

Why are there so many of these in physics?

## Accidentally correct

Sometimes in maths a mistake can lead to a correct answer. Here is a
popular example involving cancelling.

$$\frac{1\!\!\!/6}{6\!\!\!/4} = \frac{1}{4}$$

Can you find any more examples of cancelling mistakes that accidentally
give the right answer?

**Notes on this activity**

This is a bit tricky for students, but you could start by asking them
to consider fractions of the form $\frac{ab}{bc}$, as in the example.
This is essentially asking them to find solutions to the equation
$\frac{10a+b}{10b+c} = \frac{a}{c}$, which gives $10ac + bc = 10ab + ac$.
To simplify this at the outset, they could consider the case where $a = 2$.

This is now $20c + bc = 20b + 2c$, so $18c + bc = 20b$.

Rearranging for $c$ gives $c = \frac{20b}{18+b}$, and now it's just a matter of
finding a suitable value of $b$ that will result in $c$ being an integer.

Students should find that if you let $b = 6$, then $c = 5$.

Therefore $\frac{2\!\!\!/6}{6\!\!\!/5} = \frac{2}{5}$ is another example.

**Extension**

In both cases so
far, the 'cancelled'
number has been
a 6. Is this a
coincidence?

---

## Glossary

**Direct proportion**

When two quantities are in proportion using a
constant multiplier (so that an increase in one
matches an increase in the other) they are said
to be in direct proportion. It is also known as
direct variation.

**Example**

The set {12, 20, 32} is in direct proportion to
{3, 5, 8} and the constant multiplier is 4.
When interchanged, the two sets are still in
direct proportion but the constant becomes $\frac{1}{4}$.

**Inequality**

An inequality is like an equation except that it
tells you when something is less than, less than
or equal to, greater than, or greater than or
equal to something else.

**Examples**

$x > 3$ means that $x$ is greater than 3.

This inequality defines a range of possible
values for $x$, rather than a single value.

In this inequality, 3 is a lower bound for the
value of $x$.

Inequalities can also define a range with a lower
bound and an upper bound.

$1 < x < 10$ tells you that $x$ can be anything
strictly between 1 and 10.

**Inverse proportion**

When two sets of values are related and an
increase in one matches a decrease in the other,
they are said to be in inverse proportion (or
inversely proportional). This is also known as
indirect variation.

$y \propto \frac{1}{x}$ means that $y$ is inversely proportion to $x$;

or $y = \frac{k}{x}$, where $k$ is a numerical constant.

**Example**

The sets {40, 24, 15} and {3, 5, 8} are inversely
proportional since $40 \times 3 = 120$, $24 \times 5 = 120$
and $15 \times 8 = 120$.

# Extension worksheet

This worksheet develops the work on indices to include fractional indices.

Use your knowledge of the rules of indices to simplify the following.

1. $a^2 \times a^5$

2. $a^3 \times a^2 \times a^4$

3. $a^8 \times a^3 \div a^5$

4. $x^{17} \times x^{12} \div x^{15}$

5. $x^{\frac{1}{2}} \times x^{\frac{1}{2}}$

6. Simplify $\sqrt{x} \times \sqrt{x}$

7. By considering the answers to Questions **5** and **6**, write down the mathematical meaning of the index $\frac{1}{2}$.

8. Simplify $x^{\frac{1}{3}} \times x^{\frac{1}{3}} \times x^{\frac{1}{3}}$

9. Simplify $\sqrt[3]{x} \times \sqrt[3]{x} \times \sqrt[3]{x}$

10. By considering the answers to Questions **8** and **9**, write down the mathematical meaning of the index $\frac{1}{3}$.

11. $\sqrt[4]{x}$ means 'the fourth root of $x$'. Write down the equivalent index for this.

12. Write down the index representing the fifth root of $x$.

13. What does the index $\frac{1}{10}$ mean?

14. Given that $\frac{2}{3} = 2 \times \frac{1}{3}$, describe, in words, the meaning of $\frac{2}{3}$ when it is used as an index.

15. Fractional indices are generally given in the form $\frac{a}{b}$ where the denominator $b$ is the '$b$th root of …'. Write down, in words, the mathematical interpretation of the numerator $a$.

16. Simplify the following.

    a) $x^{\frac{2}{3}} \times x^{\frac{2}{3}}$                             b) $x^{\frac{3}{4}} \times x^{\frac{1}{4}} \times x^{\frac{1}{2}}$

    c) $x^{\frac{2}{5}} \times x^{\frac{4}{5}} \div x^{\frac{3}{5}}$                        d) $x^{\frac{1}{2}} \times x^{\frac{1}{3}}$

    e) $x^{\frac{2}{3}} \times x^{\frac{1}{4}} \times x^{\frac{3}{5}}$

(Answers on page 127)

# 9 Graphs

This is a very wide-ranging chapter, covering graphs of many types.

The chapter starts with a mixture of straight-line graphs and coordinate geometry.

Topics covered are:

* drawing linear graphs
* gradients of straight lines
* midpoints of line segments
* parallel and perpendicular lines.

The chapter then moves on to:

* polynomial curves, including cubics
* reciprocal curves
* exponential curves
* the square root curve
* the reciprocal of $x2$.

These graphs are also considered in practical situations, followed by work on:

* shading inequalities
* distance–time graphs
* speed–time graphs
* graphical solutions of equations.

The chapter ends with a basic introduction to calculus, finding gradients and turning points.

## Prerequisite knowledge

Students should already have a good grasp of:

* algebra
* indices
* plotting simple functions.

# Investigations, activities and puzzles

## Defining a line

If you know the gradient of a straight line and also its $y$-intercept, then you know exactly which line it is.

Are there any other ways to uniquely define a straight line?

**Notes on this activity**

Here are some ideas that students might suggest:

- two points on the line
- the two intercepts
- the gradient of the line and any point on the line

You could ask students what would be the quickest way of finding the equation of a line if you were given those pieces of information.

**Extension**

Some possible extensions are:

- Does this apply to quadratic curves?
- How many pieces of information do you need to know in order to define a quadratic curve?
- What about other kinds of curves?

This raises the question of how many pieces of information are needed to define various different mathematical objects.

Some objects require just one piece of information:

- a circle (its radius)
- a square (its side length)

As students will know, a triangle requires three pieces of information to define it:

- angle, side, angle
- side, angle, side
- side, side, side
- right-angle, hypotenuse, side

What other objects require two pieces of information to accurately define them?

One example is a rectangle. Can the students think of any others?

## Maximum box revisited

In Chapter 6, students considered the problem of finding the biggest volume of a box made by cutting squares from the corners of a big square of card and folding up the sides.

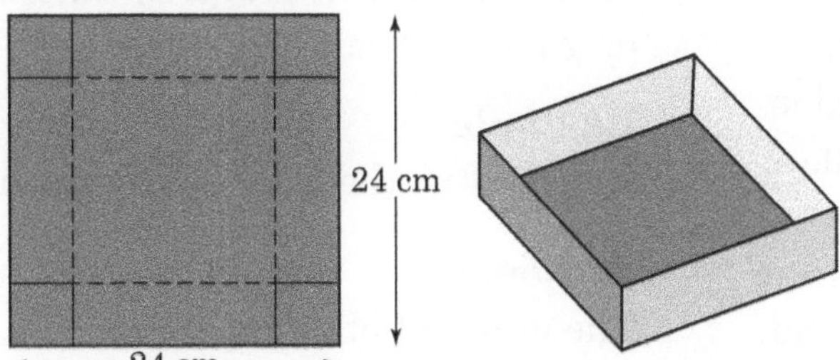


In the IGCSE™ course, practical applications of calculus are not specifically considered, so this may be a good opportunity to give the students something of an insight into this as an enrichment activity.

How can calculus be used to solve the problem of finding the maximum volume of the box?

### Notes on this activity

If the original square of card has a side length of $s$, and the squares cut out have a side length of $x$, the volume of the box $V$ turns out to be

$$V = (s - 2x)^2 \times x = 4x^3 - 4sx^2 + s^2x$$

Differentiating this and setting the expression to zero gives the equation

$$12x^2 - 8sx + s^2 = 0$$

Putting this into the quadratic formula leads to the solutions
$$x = \frac{s}{2} \text{ or } x = \frac{s}{6}.$$
Clearly, the first solution gives the minimum volume as there will be no card left from which to make the box.

The solution for the maximum volume is the second one.
This could be confirmed using the second derivative.

### Extension

As before, you could consider what would happen if the original shape were a rectangle.

What happens if you introduce an extra variable into the situation?

What if the length were a function of the width? Does this make the calculations easier?

# Glossary

### Acceleration

The acceleration of a moving object is a measure of how its velocity is changing in relation to time. It is a vector quantity, but its direction is often ignored and the word is applied only to the speed component of the velocity. It could be positive (for speeding up) or negative (for slowing down). The SI unit of acceleration is metres per second per second, abbreviated to $m/s^2$ or $ms^{-2}$.

### Example

An object starts from rest (zero velocity) and has a steady acceleration of 3 $m/s^2$. What is its velocity after 10 seconds?

The velocity will increase by 3 m/s in every second of its travel. So its velocity after 1 second is 3 m/s, after 2 seconds it is 6 m/s, after 3 seconds is 9 m/s, and so on. After 10 seconds its velocity is 30 m/s.

### Cubic graphs

These are functions where the highest power of $x$ is 3.

### Examples

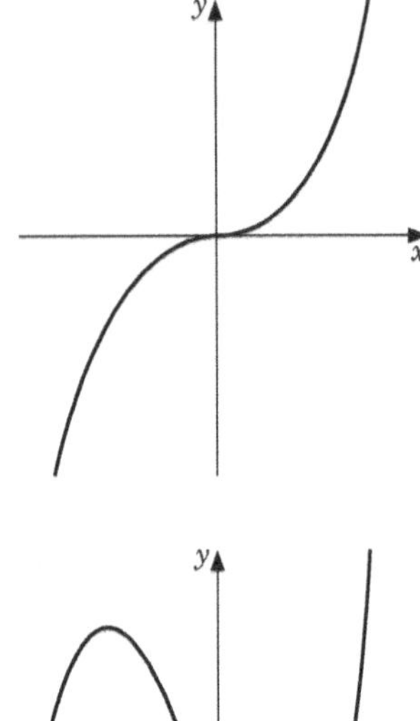

### Derivative function

The derivative function is the gradient function of any curve. It is found by differentiating the original function.

### Example

If $f(x) = 3x^2$, the gradient function $f(x) = 6x$

### Gradient

The gradient of a line drawn on a graph is a measure of its slope relative to the $x$-axis. This is expressed as the ratio of its vertical change to its horizontal change, both changes being measured on the scales of their respective axes.

### Example

In this diagram the gradient is given by $a \div b$ and is positive since $y$ increases as $x$ increases.

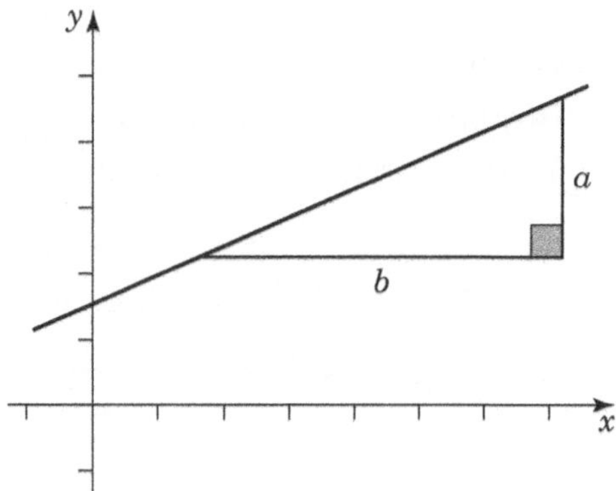

### Exponential graphs

These are functions of the form $y = a^x$

They have only one branch and one horizontal asymptote.

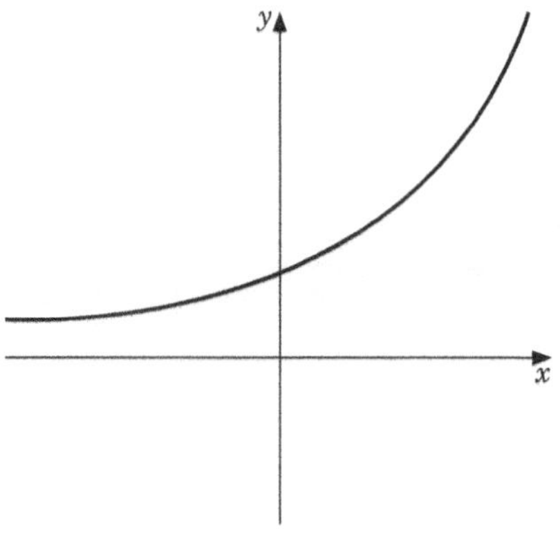

## Inequality

An inequality is shown by an expression such as $y < 3 - x$, which means that $y$ can take any value that is less than that of $3 - x$. This is shown on a graph by drawing the line $y = 3 - x$ and shading the region where $y$ is always less than the values on the line.

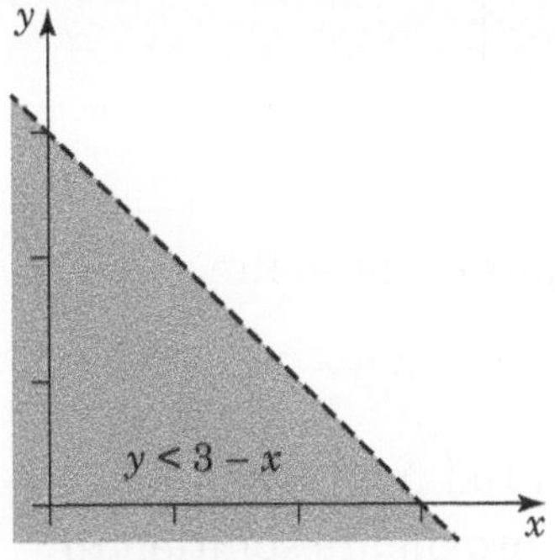


## Intercept

The intercept of a graph is the point at which it cuts across an axis.

### Example

In the linear graph shown below, the $y$-intercept is $-2$.

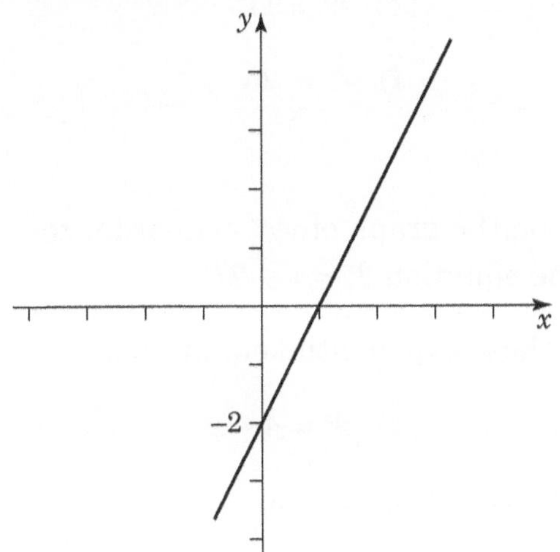


## Quadratic graph

A quadratic graph is a graph in which the relationship between the variables is given by a quadratic equation. Its shape is that of a parabola.

### Examples

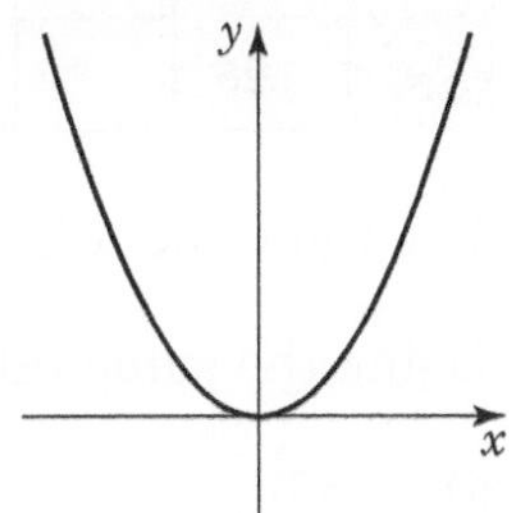

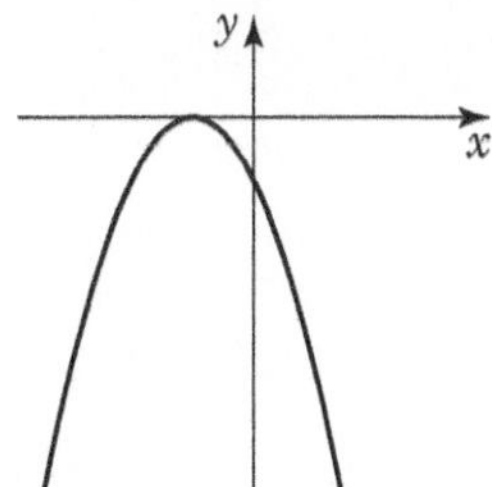

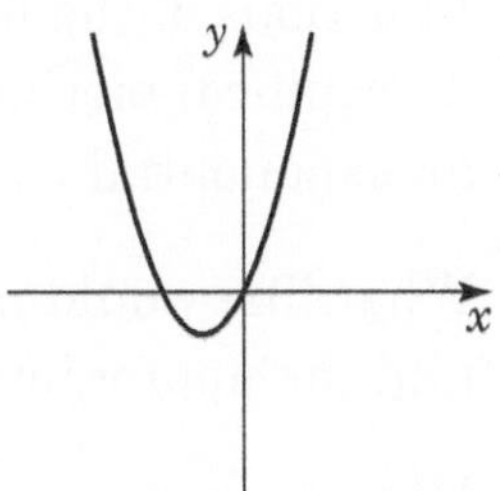

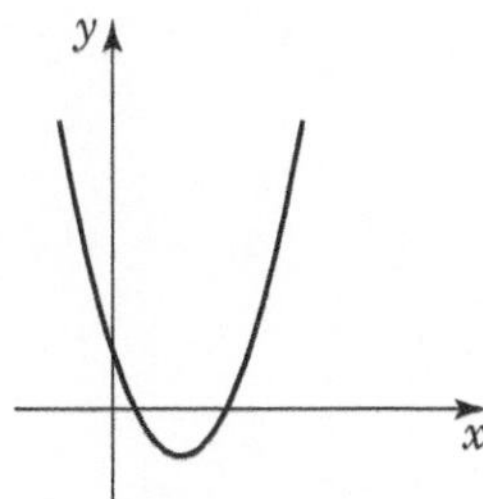


## Reciprocal graphs

These are functions of the form $y = \dfrac{x}{n} + n$

They have two branches and two asymptotes, one horizontal and one vertical.

### Example

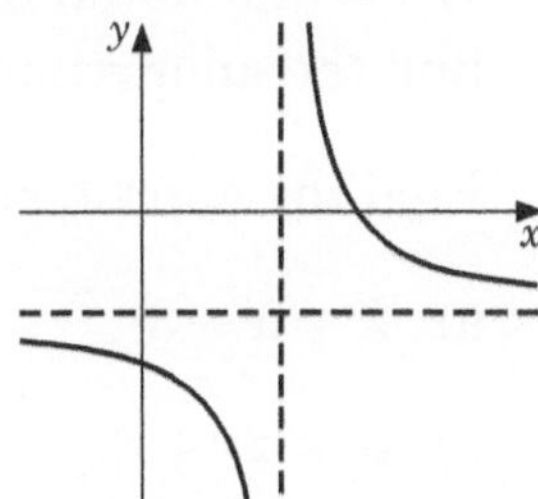


## Square root graphs

These are functions of the form $y = a\sqrt{x}$

### Example

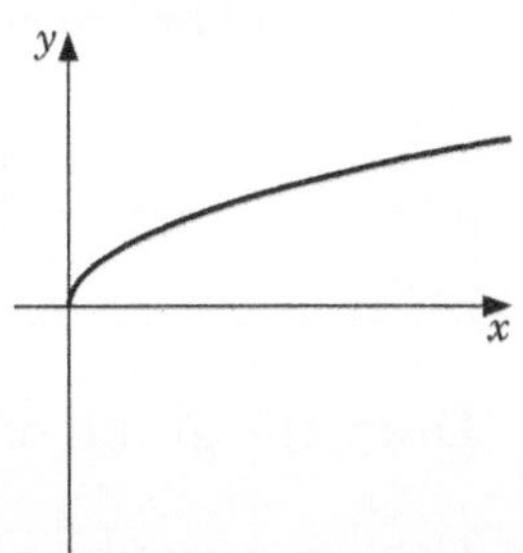

## Extension worksheet

This worksheet introduces exponential relationships of the form $y = a^x$.

1.  Copy and complete the following table of values for the function $y = 2^x$.

| $x$ | −2 | −1 | 0 | 1 | 2 | 3 | 4 |
|---|---|---|---|---|---|---|---|
| $y$ | 0.25 | | | 2 | 4 | | |

2.  Use the table to draw a graph of the function.

3.  Using the same table of values, draw the graphs of these functions.

    **a)** $y = 3^x$  **b)** $y = 4^x$  **c)** $y = 0.5^x$

    Functions of the form $y = a^x$ are called exponential functions.
    A graphical method can be used to find the approximate solution to
    an exponential equation such as $2^x = 5$.

4.  What line would need to be drawn on the graph of $y = 2^x$ in order to
    find the approximate solution to the equation $2^x = 5$?

5.  Use your graphs to find the approximate solutions to these
    exponential equations.

    **a)** $3^x = 5$  **b)** $2^x = 7$  **c)** $3^x = 11$

    **d)** $4^x = 20$  **e)** $2^x = 3.5$  **f)** $4^x = 25$

    **g)** $0.5^x = 0.1$  **h)** $3^x = 7.2$

6.  What line would need to be drawn on the graph of $y = 2^x$ in order to
    find the approximate solution to the equation $2^x = x + 2$?

7.  Find the approximate solutions to these exponential equations.

    **a)** $2^x = x + 3$  **b)** $3^x = 2 - x$  **c)** $2^x = 4 - x$

    **d)** $3^x = 2x + 3$  **e)** $4^x = 5 - 0.5x$  **f)** $2^x = x^2$

8.  The exponential function is given by the formula $y = e^x$ where the
    letter e stands for a number (e $\approx 2.718$). Using the $e^x$ button on your
    calculator, or the approximate value for e given above, copy and
    complete the following table of values (round to three significant
    figures where necessary):

| $x$ | −2 | −1 | 0 | 1 | 2 | 3 | 4 |
|---|---|---|---|---|---|---|---|
| $y$ | | | 1 | 2.72 | | | |

9.  Draw the graph of $y = e^x$.

10. Use your graph to find the approximate solution to the equation $e^x = 2$.

(Answers on page 128)

This is a short chapter that builds on the ideas studied in Trigonometry 1.

Whereas the focus of the earlier chapter was on right-angled triangles, this chapter is about trigonometry in any triangle.

The chapter opens by looking at the graphs of the three main trigonometric functions, and considering angles greater than 180°.

Students then go on to learn the sine and cosine rules.

The chapter closes with an exercise applying these ideas to bearings.

## Prerequisite knowledge

This chapter assumes that students have a basic understanding of:

- trigonometry in right-angled triangles
- bearings
- basic angle facts.

# Investigations, activities and puzzles

## Heron's formula

An interesting enrichment here could be to consider the different formulae for finding the area of a triangle.

In addition to the formula area $= \frac{1}{2} \times$ base $\times$ height the trigonometric formula area $= \frac{1}{2}ab\sin C$ can be used when two side lengths and the size of the angle between them are given.

If the three side lengths are given, Heron's formula can be used:

$$A = \sqrt{s(s-a)(s-b)(s-c)}$$

where $s$ is half of the perimeter of the triangle.

Is there a formula for the area of a triangle that uses two angles and the length of the side between them?

**Notes on this activity**

The underlying idea here is that if you have enough information about something to be able to define it, then you should be able to use that information to work out anything else you need to know about that object. This idea is similar to the 'defining a line' activity in the previous chapter.

**Extension**

Heron's formula for the area of a triangle is a consequence of Brahmagupta's formula for the area of a cyclic quadrilateral, since all triangles can be thought of as cyclic quadrilaterals with one side of length zero. Students could be asked to consider why all triangles are cyclic but not all quadrilaterals are cyclic.

## Odd and even functions

The function $y = \cos x$ is an even function as it has a line of symmetry along the $y$-axis.

The function $y = \sin x$, however, does not. It is called an odd function because it has rotational symmetry of order 2 about the origin.

How many other odd and even functions can you think of?

## The longest side

The longest side of a triangle is always opposite the biggest angle. Why is this?

## Maps

Bearings are important for ships navigating at sea, or planes flying in the air.

Over the centuries, different maps have been created to help people to find their way around.

The most popular one is Mercator's map from 1569. It became popular because although some things are distorted by the process of turning the surface of a sphere into a rectangle, one thing that is kept accurate on this particular map is all of the bearings that you would need to travel on to get from one place to another.

Investigate other map projections.

What are their strengths and weaknesses?

**Notes on this activity**

Bearings are problematic for many students, so explaining why they exist, along with a bit about their history, could be beneficial.

There are many different map projections, some of which they may be already familiar with.

Some well-known alternative projections are:

- the Gall–Peters projection
- the Robinson projection.

**Extension**

This could be extended for the more advanced students to consider some of the mathematics involved in map projections.

---

## Glossary

### Cosine rule

The cosine rule is an extension of Pythagoras' theorem, which can be applied to any triangle.

$$a^2 = b^2 + c^2 - 2bc \cos A \quad \text{or} \quad \cos A = \frac{b^2 + c^2 - a^2}{2bc}$$

$$b^2 = a^2 + c^2 - 2ac \cos B$$

$$c^2 = a^2 + b^2 - 2ab \cos C$$

### Sine rule

The sine rule is based on the fact that in any triangle the length of a side is proportional to the sine of the angle opposite that side.

$$\frac{a}{\sin A} = \frac{b}{\sin B} = \frac{c}{\sin C}$$

## Extension worksheet

This worksheet looks at some fundamental trigonometric identities.

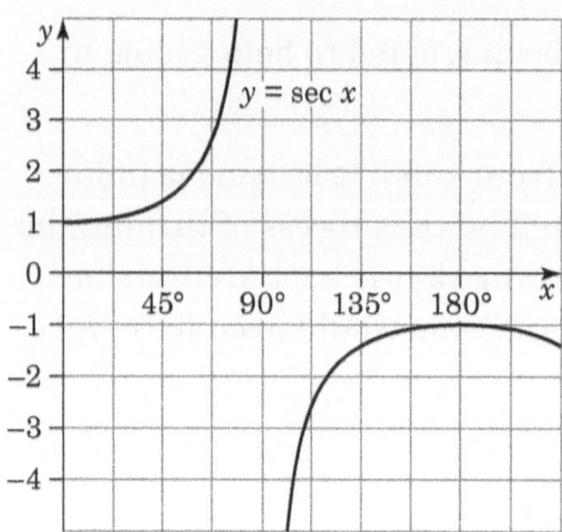

In the right-angled triangle, $\sin x = \dfrac{a}{c}$, $\cos x = \dfrac{b}{c}$ and $\tan x = \dfrac{a}{b}$.

1.  Use the formulae given above to show that $\dfrac{\sin x}{\cos x} = \tan x$.

2.  By writing expressions for $\sin^2 x$ and $\cos^2 x$, use Pythagoras' theorem to show that $\sin^2 x + \cos^2 x = 1$.

    The secant of $x$ (sec $x$) is defined as $\dfrac{1}{\cos x}$ and the cosecant of $x$ (cosec $x$) is defined as $\dfrac{1}{\sin x}$.

3.  Write expressions for sec $x$ and cosec $x$ in terms of $a$, $b$ and $c$.

4.  Write a similar expression for the cotangent of $x$ (cot $x$).

5.  Use these definitions to show that:

    **a)**  $1 + \tan^2 x = \sec^2 x$       **b)**  $1 + \cot^2 x = \operatorname{cosec}^2 x$

6.  Copy and complete the table of values for sec $x$, cosec $x$ and cot $x$ (if a value is *undefined*, leave it blank).

| $x$ | 0° | 15° | 30° | 45° | 60° | 75° | 90° | 105° | 120° | 135° | 150° | 165° | 180° |
|---|---|---|---|---|---|---|---|---|---|---|---|---|---|
| **sec $x$** | | | | | | | | | | | | | |
| **cosec $x$** | | | | | | | | | | | | | |
| **cot $x$** | | | | | | | | | | | | | |

7.  Draw graphs of $y = \sec x$, $y = \operatorname{cosec} x$ and $y = \cot x$. Draw each graph on a separate set of axes and do not join points if they have an undefined value between them.

8.  The values of $x$ for which the functions are undefined create *asymptotes*. Mark on your graphs, using dotted vertical lines, the asymptotes for each function.

9.  Use your graphs to find approximate solutions to the following trigonometric equations.

    **a)**  $\cot x = 2$     **b)**  $\operatorname{cosec} x = 1.5$     **c)**  $\sec x = -2.5$

(Answers on page 129)

This is another short chapter that is split into two main sections.

The first section covers Venn diagrams and set theory, in detail.

Students are required to construct and interpret Venn diagrams containing two or three sets. Technical language is used throughout this section, such as the words:

- intersection
- union
- complement
- element
- universal set
- subset
- empty set.

The second section covers the topic of functions. This includes:

- flow diagrams
- domain and range
- composite functions
- inverse functions.

## Prerequisite knowledge

This chapter assumes that students can already:

- rearrange a formula
- work with basic inequalities
- substitute values into an algebraic expression.

# Investigations, activities and puzzles

## De Morgan's laws

Ask students to shade in the areas on the Venn diagrams provided on the next page.

What do they notice?

Can they work out any rules based on what they have discovered?

**Notes on this activity**

At some point in the process, students should notice that the diagrams in the second half of the sheet are the same as the diagrams in the first half.

Once they have discovered all eight pairs of matching diagrams, they could begin to think about why this is the case.

It should be noted that students generally find this activity quite difficult, so some time should be spent at the beginning thoroughly going over the ideas of union, intersection and complement.

Time should also be spent explaining the best way to shade these regions.

It is a good idea to print some spare sheets as some students are likely to get some questions wrong and ask for an extra copy.

Students should end up discovering De Morgan's laws.

These say that

$$(A \cup B)' = A' \cap B'$$

and

$$(A \cap B)' = A' \cup B'$$

This means that if students are asked to shade the complement of a bracketed expression, they can remove the brackets as long as they find the complement everything inside the brackets, and change a union into an intersection and vice versa.

**Extension**

Can these ideas be extended to Venn diagrams containing more than two sets?

Shade in the following sets according to their descriptions.

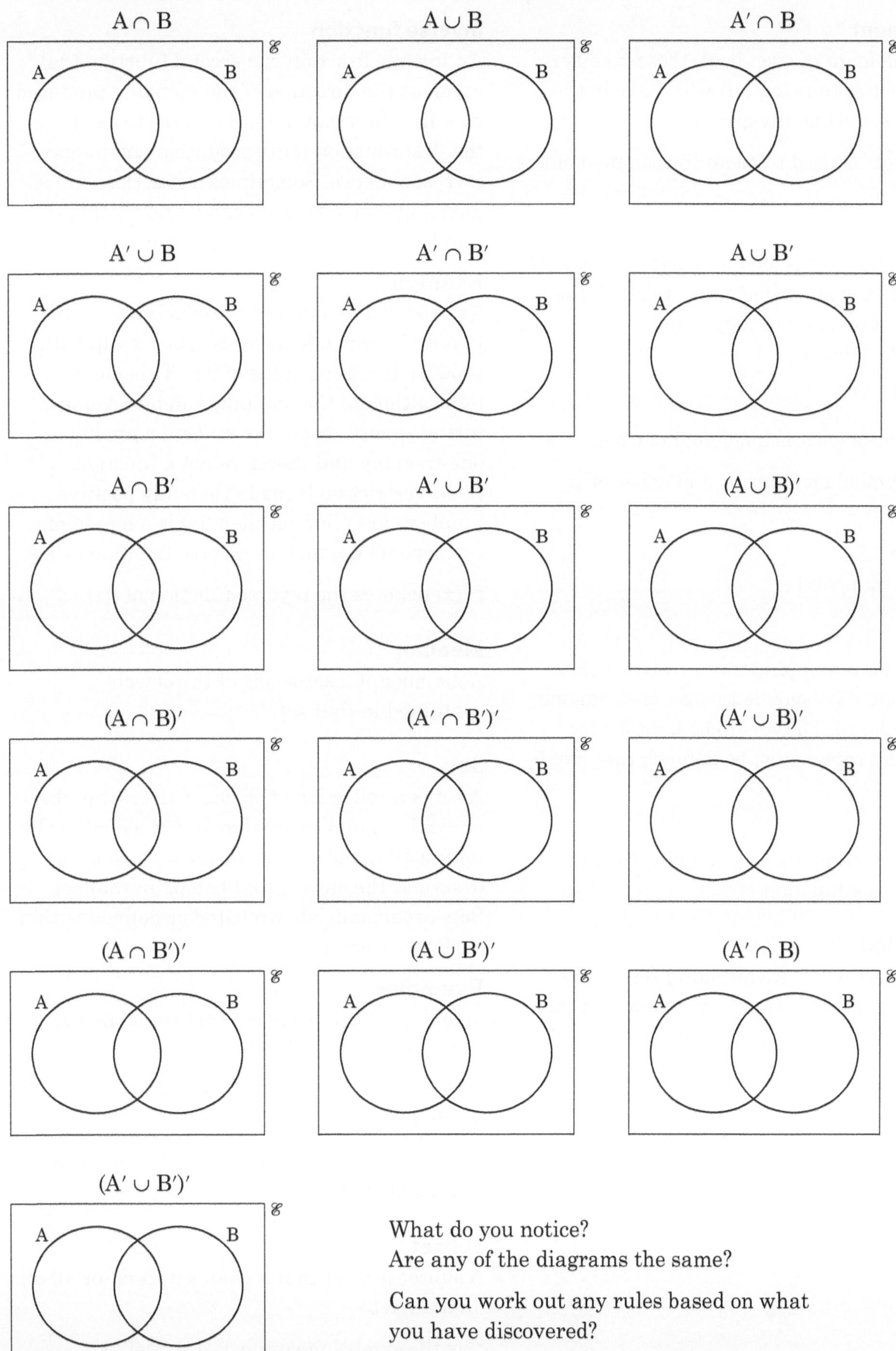


What do you notice?
Are any of the diagrams the same?

Can you work out any rules based on what you have discovered?

# Glossary

## Complement

The complement of a set is all those members that are not in that set, but which are in the universal set originally given.

The symbol $'$ is used to mean the complement of a set.

### Example

If the universal set is {odd numbers less than 30} and the set A is {all prime numbers}, then the complement of A is shown by A′ and is {1, 9, 15, 21, 25, 27}.

## Element

An element of a set is a member of a set.

$\in$ is the symbol meaning 'is a member of' or 'belongs to'.

### Example

$2 \in$ {even numbers}

## Function

A function is a mapping that involves either a one-to-one correspondence or a many-to-one correspondence. The sets to be used for the domain and range must be defined; they can be the same.

### Example

For positive numbers, the mapping 'is the square of' is a function.

## Intersection

The intersection of two (or more) sets is the single set made containing only members that are common to both (or all) of the original sets.

$\cap$ is the symbol for the intersection of sets.

### Example

$\{4, 7, 13, 20\} \cap \{2, 7, 10\}$ is $\{7\}$

## Inverse function

An inverse function is a second function that reverses the direction of the mapping produced by a first function. For an inverse to exist, the first function must produce a one-to-one correspondence. Sometimes a function can be forced into being one-to-one by restricting the numbers to be used in the domain and range.

### Example

$f(x) = x^2$ is a function of $x$ producing a many-to-one correspondence since $x$ or $-x$ will both produce the same value of $f(x)$. This means that, although the mapping can be reversed (using square roots), the reverse mapping is one-to-many and therefore not a function. If the restriction is made that only positive numbers are allowed, then $f(x)$ is a one-to-one correspondence and an inverse function exists.

$f^{-1}(x)$ denotes the inverse function of $f(x)$.

## Member

A member of a set is one of the objects contained in that set.

## Set

A set is a collection of objects (letters, numbers or symbols, and so on) that is defined either by listing all the objects or by giving a rule that describes the objects that belong in that set. Sets are usually shown listed or defined within curly brackets: { }.

### Examples

{a, e, i, o, u} is a listed set that could also be described by the rule {the vowels}

{5, a person, a table, Z} is a listed set for which a rule would be difficult to find.

{all the numbers} is a rule for a set that it is impossible to list.

## Subset

A subset is a set that contains part of (or all of) another set.

$\subset$ is the symbol meaning 'is a subset of'.

### Example

$\{2, 7, f, t, M, \varphi\} \subset$ {numbers, letters, symbols}

## Union

The union of two (or more) sets is their combination into a single set containing all the members of the original sets. A member found in more than one of the original sets need only be shown once in the union.

$\cup$ is the symbol for the union of sets.

### Example

$\{4, 7, 13, 20\} \cup \{2, 7, 10\}$ is $\{2, 4, 7, 10, 13, 20\}$

## Universal set

A universal set is a set that is first defined (by list or rule) and within which all the statements that follow must be interpreted.

### Examples

After giving the universal set {positive numbers less than 10}, the set {even numbers} would be only $\{2, 4, 6, 8\}$.

In the universal set {all positive numbers}, $x^2 = 4$ has only the solution $x = 2$ since $-2$ is not in the universal set.

$\mathscr{E}$ is a symbol for the universal set, but others are also commonly used.

## Venn diagrams

Venn diagrams are used to give a pictorial view of the relationships of sets and subsets within a universal set; the universal set is shown enclosed by a rectangle, and all others by circles or simple closed curves.

### Examples

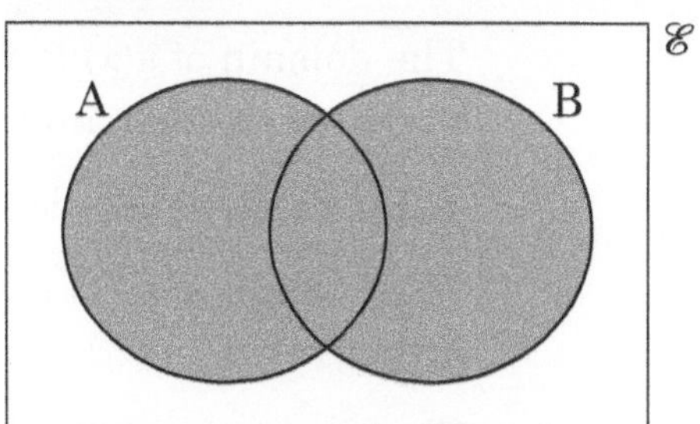

$A \cup B$
union of A and B

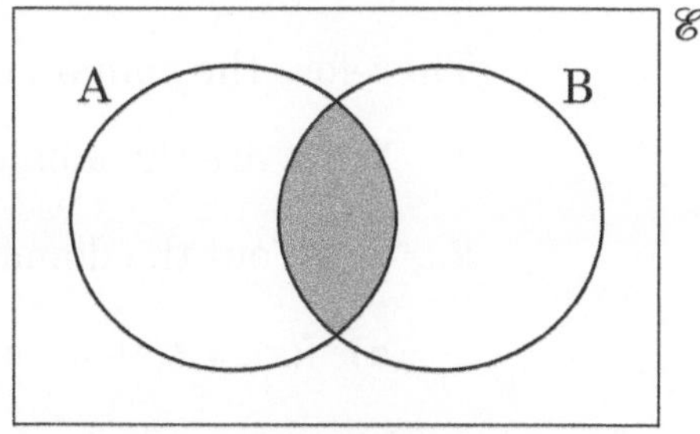

$A \cap B$
intersection of A and B

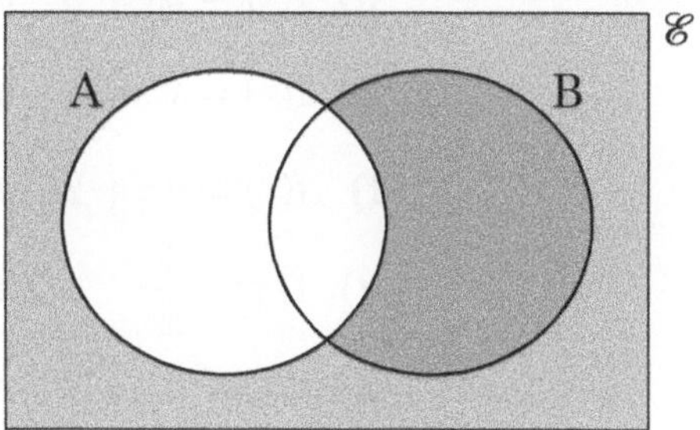

$A' =$ complement of A

## Extension worksheet

This worksheet develops the work on functions by considering the range and domain of different composite functions.

Consider the two functions $f(x) = x^2 + 1$ and $g(x) = \sqrt{x}$.

The domain of $f(x)$ is any real number (we write this $x \in \mathbb{R}$) but the range is $f(x) \geqslant 1$.

The domain of $g(x)$ is $x \geqslant 0$, since you can't take the square root of a negative number, and the range is $g(x) \geqslant 0$.

What then are the domain and range of $fg(x)$ and $gf(x)$?

The domain of $fg(x)$ is $x \geqslant 0$ since these are the numbers that can go into $g(x)$.

The range of $g(x)$ then become the domain of $f(x)$ within the composite function.

If we restrict the domain of $f(x)$ to $x \geqslant 0$, this will restrict the range of $fg(x)$ to $fg(x) \geqslant 1$.

Therefore the domain of $fg(x)$ is $x \geqslant 0$ and the range is $fg(x) \geqslant 1$.

1.  What are the domain and range of $gf(x)$?

2.  Work out the domain and range of $fg(x)$ when:

    **a)** $f(x) = 2x + 3$    and    $g(x) = x^2 - 1$

    **b)** $f(x) = 3x - 2$    and    $g(x) = \sqrt{x}$

    **c)** $f(x) = 4x + 2$    and    $g(x) = \dfrac{1}{x}$

    **d)** $f(x) = 2x + 3$    and    $g(x) = x^2 + 2$

    **e)** $f(x) = x^2 + 1$    and    $g(x) = \dfrac{1}{x - 3}$

    **f)** $f(x) = 2x + 2$    and    $g(x) = 2^x$

    **g)** $f(x) = \dfrac{1}{x}$    and    $g(x) = 1 + \sqrt{x}$

    **h)** $f(x) = 5x - 7$    and    $g(x) = x^3$

(Answers on page 130)

This chapter is split into two related sections.

The first section covers vectors.

Students learn what a vector is and the various different ways there are of writing a vector. They then learn how to:

- add vectors
- subtract vectors
- multiply a vector by a scalar
- recognise parallel vectors
- find the magnitude of a vector
- solve geometrical problems using vectors.

The second section deals with transformations. The four main transformation types are studied:

- reflection
- rotation
- translation
- enlargement

The chapter closes with a look at how transformations can be combined.

### Prerequisite knowledge

This chapter assumes that students are already familiar with:

- Cartesian coordinates
- Pythagoras' theorem
- negative numbers
- the idea of a scale factor.

## Investigations, activities and puzzles

### Combining transformations

Often the order in which transformations occur makes a difference.

For example, consider the following diagrams.

In this diagram, triangle A has been reflected in the $y$-axis, and then reflected in the line $y = x$ to make triangle A′.

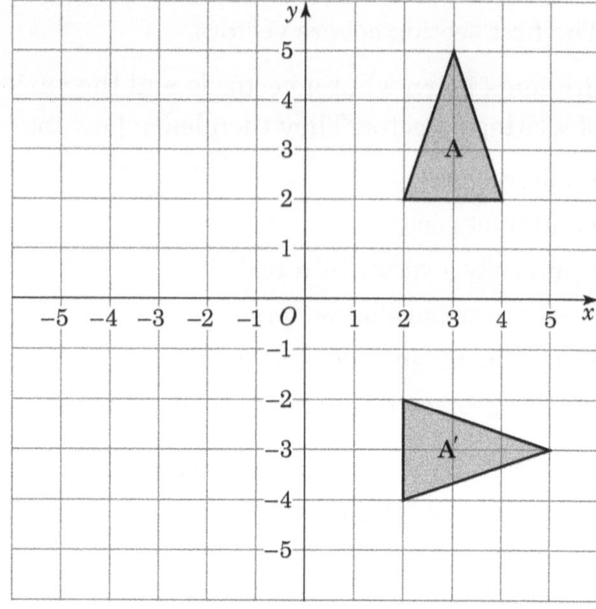

In this second diagram, triangle A has been reflected in the line $y = x$, and then reflected in the $y$-axis to make triangle A′.

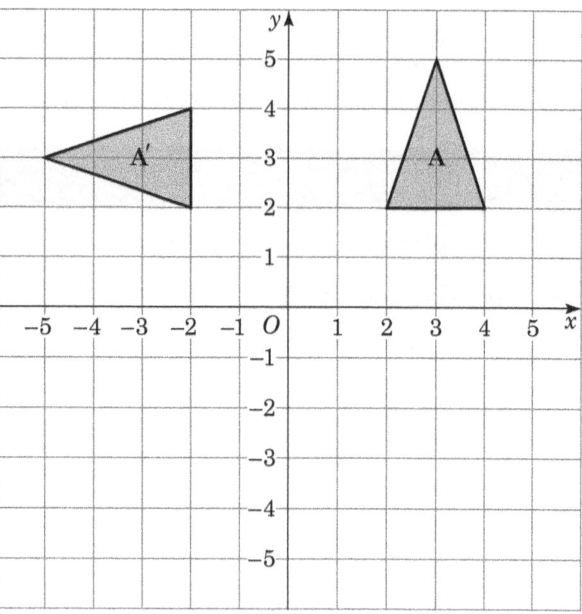

As you can see, performing the transformations in a different order leads to different outcomes.

Experiment with transforming shapes. Does the order always matter?

Can you find two transformations that give the same result regardless of the order in which you perform them?

### Extension

If you want to, you could extend this to consider simple ideas related to group theory.

### Notes on this activity

Students should be encouraged to consider reflections in the $x$-axis, the $y$-axis, the line $y = x$ and the line $y = -x$.

They can also consider rotations around the origin in multiples of 90°.

## Enlarging a picture

Find a picture online or from a magazine and enlarge it.

Follow these steps:

- Decide on a scale factor based on how large you want to make the new picture.
- Measure the distances to significant points on the original picture.
- Multiply these distances by your chosen scale factor.
- Plot the new points.
- Join together the new points to make your enlarged picture.

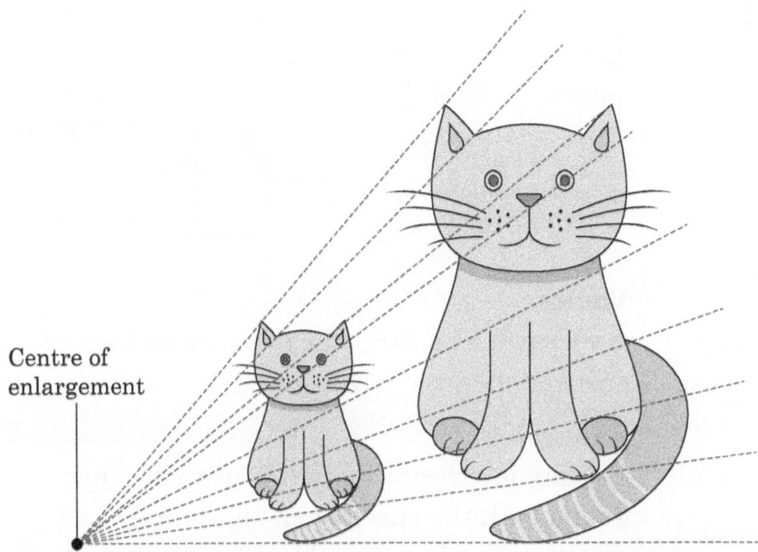


**Notes on this activity**

Students need to carefully consider what their scale factor of enlargement is going to be, based on the size of the paper they are working on. They also need to carefully consider where they are going to put their centre of enlargement, as it will determine how the two pictures are situated relative to each other.

# Glossary

## Enlargement

An enlargement is a transformation by which the distances between every pair of points on the object are multiplied by the same amount to produce a different-sized image of the object. The multiplier can be any non-zero whole number or fraction.

### Examples

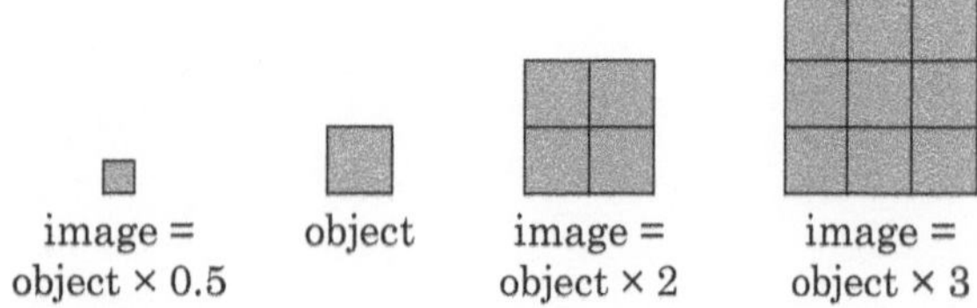


## Modulus

The modulus of a vector is a measure of its length when it is represented by a single straight line and is the positive square root of the sum of the squares of the components of the vector. It is also known as the magnitude.

## Reflection

A reflection is a transformation such that any two corresponding points on the object, and the image, are both the same distance from a fixed straight line, and a line drawn between those points is perpendicular to that fixed line.

It is described by giving the position of the fixed line.

## Rotation

A rotation is a transformation about a fixed point such that every point on the object turns through the same angle relative to that fixed point.

A rotation is described by giving the angle and direction of the turn, and the position of the fixed point about which the turn is made.

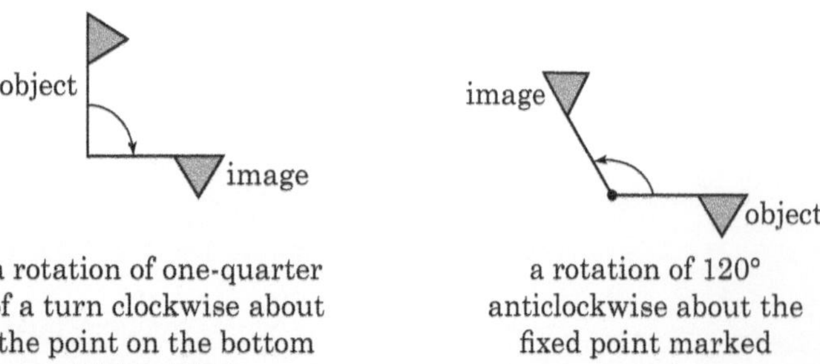


a rotation of one-quarter
of a turn clockwise about
the point on the bottom
of the object

a rotation of 120°
anticlockwise about the
fixed point marked

## Translation

A translation is a transformation such that every point on the object can be joined to its corresponding point on the image, by a set of straight lines that are all parallel and of equal length. A translation is described by the direction and length of the movement.

### Examples

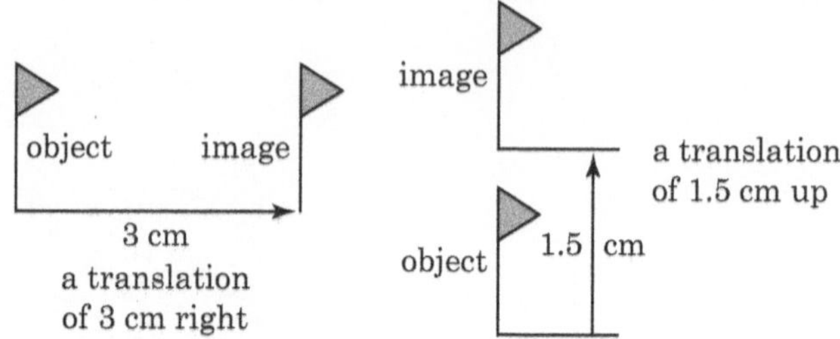


## Vector

A vector can be defined by two quantities: its size and its direction.

### Examples

Velocity is a vector since it is described by giving both the speed of an object and the direction in which the object is moving (speed alone is not a vector). Force is also a vector.

# Extension worksheet

This worksheet develops the work on vectors to include vectors in three dimensions.

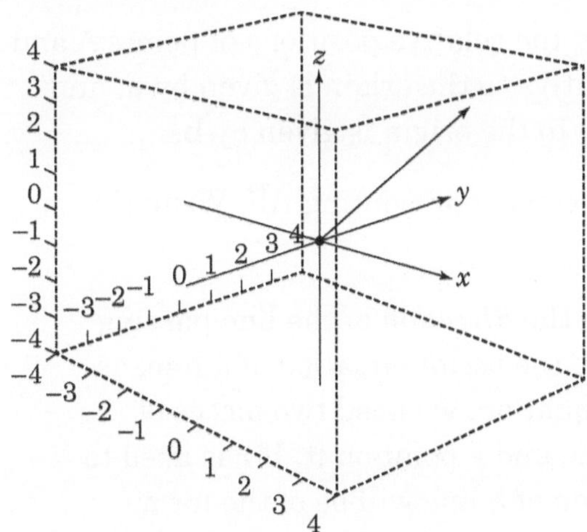

This diagram shows a vector in three dimensions, $x$, $y$ and $z$.

It can be described as a column vector, in this case $\begin{pmatrix} 2 \\ 1 \\ 3 \end{pmatrix}$.

More generally, any vector in three dimensions can be written as $\begin{pmatrix} x \\ y \\ z \end{pmatrix}$.

Vectors in three dimensions behave, arithmetically, the same as those in two dimensions.

$$\mathbf{a} = \begin{pmatrix} 2 \\ 1 \\ 3 \end{pmatrix}, \mathbf{b} = \begin{pmatrix} -1 \\ 2 \\ 4 \end{pmatrix} \text{ and } \mathbf{c} = \begin{pmatrix} 3 \\ -2 \\ -1 \end{pmatrix}$$

1. Using the vectors on the previous page, find the following vector sums.

   **a)** $\mathbf{a} + \mathbf{b}$      **b)** $2\mathbf{a}$      **c)** $\mathbf{a} + 2\mathbf{c}$      **d)** $2\mathbf{b} + 3\mathbf{c}$

   **e)** $\mathbf{a} + \mathbf{b} + \mathbf{c}$      **f)** $3\mathbf{a} + 4\mathbf{c}$      **g)** $\mathbf{a} - \mathbf{b}$      **h)** $2\mathbf{b} - 3\mathbf{a}$

2. Draw a rough diagram showing the relative positions of points A and B if the position vector of A relative to the origin is given by $\mathbf{a}$, and the position vector of B relative to the origin is given by $\mathbf{b}$.

3. Find, in terms of $\mathbf{a}$ and $\mathbf{b}$, the vector representing $\overrightarrow{AB}$. Write this as a column vector.

The vector $\overrightarrow{AB}$ can be thought of as the *direction* of the line passing through points A and B. To describe the *vector* equation of a line, as opposed to the normal Cartesian equation, we need two pieces of information: the direction of the line and a point on it. If $\mathbf{r}$ is used to describe the line, the vector equation of a line will be of the form

$$\mathbf{r} = \overrightarrow{OA} + \lambda\overrightarrow{AB} \text{ where } \lambda \text{ is a scalar parameter.}$$

4. Write down the vector equation of the line passing through A and B.

5. Write down the coordinates of the point on the line given by the following values of $\lambda$.

   **a)** $\lambda = 1$      **b)** $\lambda = 2$      **c)** $\lambda = -1$      **d)** $\lambda = -3.5$

6. For the points A(4, 5, 2) and B(−1, 6, 8) write down:

   **a)** the vectors $\overrightarrow{OA}$ and $\overrightarrow{OB}$

   **b)** the vector $\overrightarrow{AB}$

   **c)** the vector equation of the line passing through A and B.

(Answers on page 130)

The book now turns to look at data.

The chapter starts with a few basic graphs:

- pictograms
- tally charts
- pie charts
- two-way tables
- dual bar charts
- composite bar charts

Students then move on to consider the three averages and the range of a set of data.

There is a substantial exercise on this topic that also covers finding the mean and median from a frequency table.

Then students look at how data can be presented using:

- frequency charts
- stem-and-leaf diagrams
- back-to-back stem-and-leaf diagrams
- frequency polygons
- histograms.

Students then study scatter diagrams, along with the ideas of correlation and a line of best fit.

The chapter closes by considering how a student might compare two sets of data.

## Prerequisite knowledge

Before attempting this chapter, students should be able to:

- read a scale accurately
- use a calculator
- measure an angle using a protractor
- calculate fractions of quantities.

# Investigations, activities and puzzles

## An average puzzle

Can you find 5 numbers with a mean of 4, a median of 2 and a mode of 1?

An answer to this is 1, 1, 2, 5, 11.

Can you find another set of 5 numbers with the same set of averages?

**Notes on this activity**

This is a good way to get students to think about different kinds of averages. Students also experiment with sets of data that have no mode or are bimodal.

**Extension**

You can extend this activity by considering the following:

- Are you allowed to use the same number more than once?
- Are you allowed to use zero?
- Are you allowed to use negative numbers?
- Are there any values for the three averages for which you can't find a set of numbers? Why is this?
- Are there any values for the three averages for which the set of numbers is unique?
- What happens if you increase the size of the set of numbers?
- What happens to the averages if you:
  - multiply all the numbers in the list by 2?
  - add the same amount to each of the numbers?

## How many marbles are in the jar?

Ask students to estimate how many marbles there are in a jar.

When you average out all of their guesses, how accurate is the estimate?

Does the estimate become more accurate the more guesses you have?

**Notes on this activity**

This activity is based on the idea of 'the wisdom of crowds', which says that large groups of people are more intelligent than individuals.

Does this seem to be true in this case?

## Average test scores

A student takes two tests. In the first test they score 60% and in the second test they score 90%. Their overall average score is recorded on their report card as 75%, since 75 is the average of 60 and 90.

The student complains and says that they have looked at the two tests and discovered that they actually got 80% of the questions correct.

How has this happened?

**Notes on this activity**

Many students will not know that it is a bad idea to average percentages.

One way that the situation described could have arisen was if in the first test they scored 30 out of 50, and in the second test they scored 90 out of 100. The problem is brought about by the fact that the two tests are marked out of two different totals.

This is exactly the same issue that we saw in the 'A trip to the seaside' puzzle that we looked at in Chapter 3. The incorrect answer of 25 km/h was found by averaging the two different average speeds.

## Florence Nightingale

Florence Nightingale is mentioned in the introduction to this chapter in the Student Book.

Students could do some research and find out about the contributions she made to statistics.

## Lengths of words

Find an author who has written books for children, and also books for adults. This will be easy as many authors do this.

Look at the first 100 words in one of their adult books and the first 100 words in one of their children's books.

Do they use longer words in the books for adults?

**Notes on this activity**

The results can be made into a composite bar chart.

Alternatively, rather than using two different books written by the same author, you could look at how the same news story has been reported in several different kinds of newspaper.

# Glossary

## Bar chart

A bar chart is a frequency diagram using rectangles of equal width whose heights or lengths are proportional to the frequency.

Usually, adjacent rectangles or bars only touch each other if the data is continuous; for discrete data a space is left between the bars. The bars may be of any width and sometimes are no more than lines.

### Example

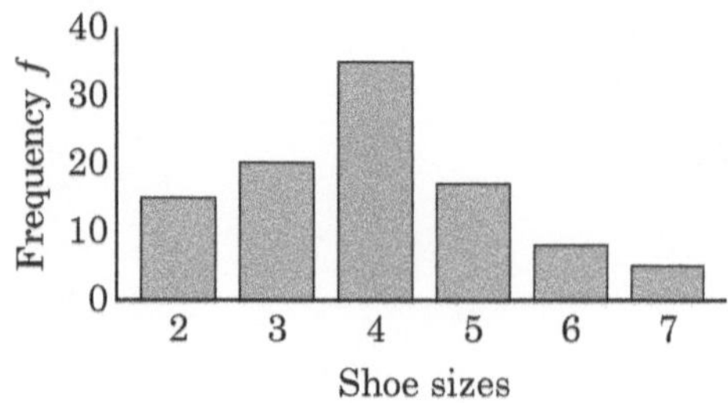


## Cumulative frequency

Cumulative frequency is the total of all the frequencies of a set of data up to any piece or group of data.

## Histogram

A histogram is a frequency diagram that consists of rectangles whose widths are proportional to the class interval and whose areas are proportional to the frequency.

The class intervals may or may not be of equal width; if they are of equal width, then the histogram is indistinguishable from a bar chart.

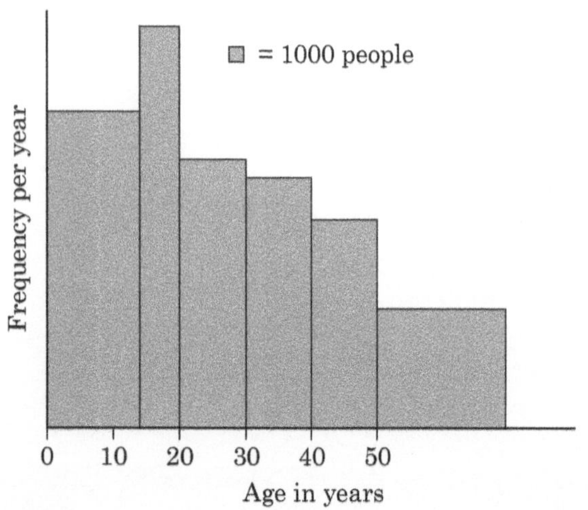


## Interquartile range

The interquartile range of a set of data is the difference in value between the lower and upper quartiles for that data. It is one way of measuring the dispersion (spread) of the data.

## Mean

The mean value of a set of data is usually taken to be the arithmetic average.

## Median

The median value of a set of data is the numerical value of the piece of data in the middle of the set, after arranging the set in size order. If there is an even number of pieces of data, the mean of the middle two is taken as the median.

### Example

Data 9, 3, 3, 15, 11 has a median of 9 (middle of 3, 3, 9, 11, 15) whereas 6, 2, 12, 4, 7, 18 has a median of 6.5 (mean of middle pair 6 and 7).

## Mode

The mode of a set of data is the piece of data found most often.

### Example

Data 9, 3, 3, 15, 11 has a mode of 3 (as 3 is the most common number found in the list of data).

## Pictogram

A pictogram is a frequency diagram using a symbol to represent a particular number of units of data. The symbol usually relates to the data being shown.

### Example

In this pictogram, the $ symbol represents $1.

| Reena | $ $ |
| --- | --- |
| Sharon | $ $ $ $ |
| Tim | $ $ |
| June | $ $ $ |

## Pie chart

A pie chart is a circular frequency diagram using sectors whose angles at the centre are proportional to the frequency.

### Example

This pie chart shows the different items sold at a garden centre.

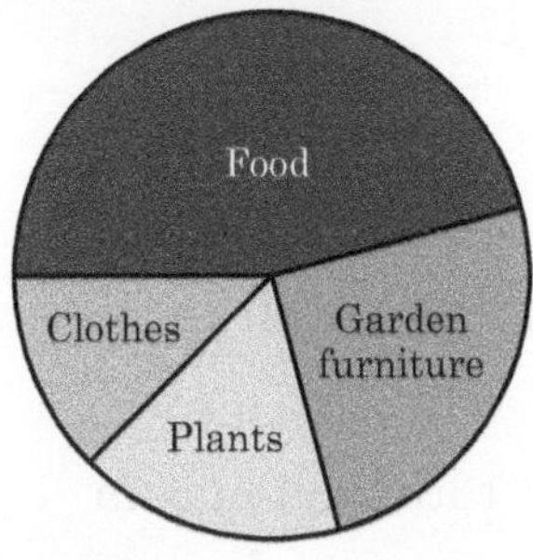


## Range

The range of a set of numerical data is the numerical difference between the smallest and the greatest values to be found in that data.

### Example

For the data 9, 3, 3, 15, 11, the range is $15 - 3 = 12$.

## Scatter diagram

A scatter diagram shows how two sets of numerical data are related, by treating matching pairs of numbers as coordinates and plotting them as a single point, and then repeating this as necessary for each data-pair.

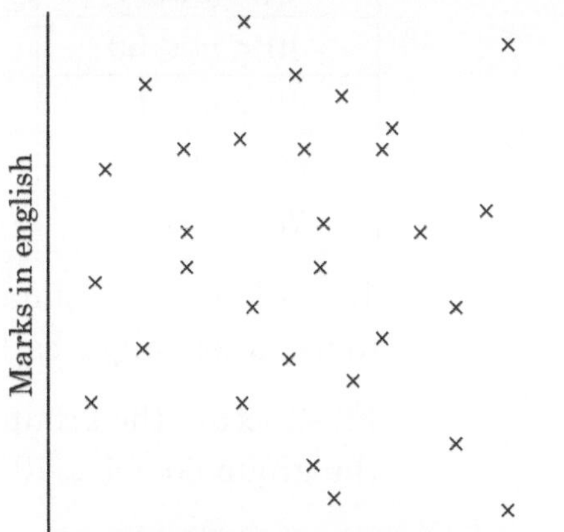


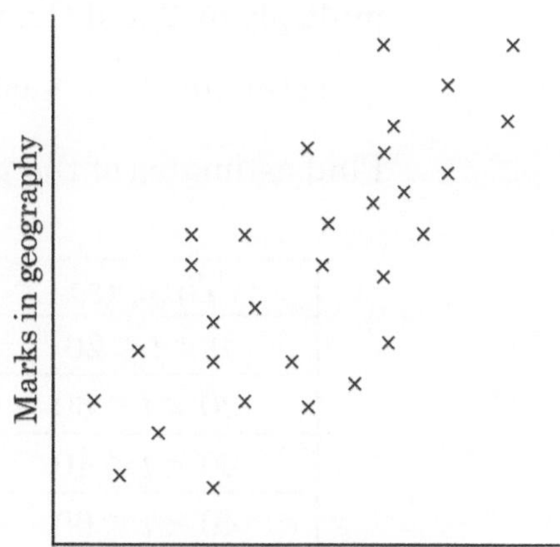


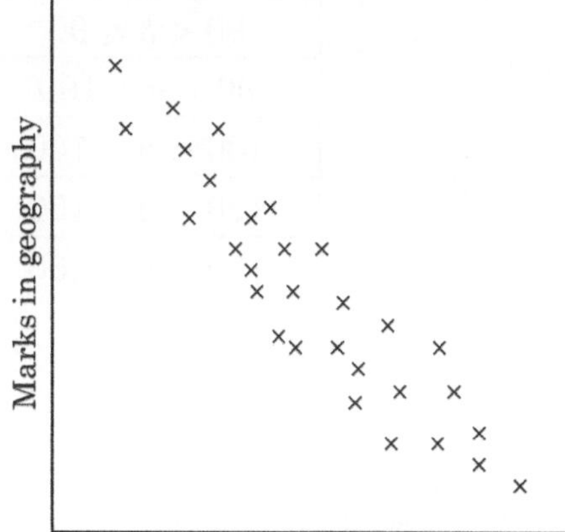

# Extension worksheet

This worksheet develops the work on grouped data and looks at using linear interpolation to work out an estimate of the median.

The grouped frequency table shows the masses of 25 people.

| Mass (kg) | Frequency |
|---|---|
| $40 < w \leqslant 50$ | 2 |
| $50 < w \leqslant 60$ | 9 |
| $60 < w \leqslant 70$ | 11 |
| $70 < w \leqslant 80$ | 3 |

To work out an estimate for the median mass (the 13th person), you need to use a technique called *linear interpolation*.

First, locate the group that contains the median: the 13th person is in the group $60 < w \leqslant 70$.

You then divide the width of the class by the number of people in that class ($10 \div 11$). The 13th person is the second in that class so you multiply by 2 and then add the starting value of the class.

$$\text{estimate of the median} = \frac{10}{11} \times 2 + 60 = 61.8 \ (3 \text{ s.f.})$$

Find estimates of the median for the following sets of data.

1.

| Time (s) | Frequency |
|---|---|
| $0 < t \leqslant 20$ | 5 |
| $20 < t \leqslant 30$ | 15 |
| $30 < t \leqslant 40$ | 8 |
| $40 < t \leqslant 60$ | 3 |

2.

| Height (cm) | Frequency |
|---|---|
| $80 < h \leqslant 90$ | 6 |
| $90 < h \leqslant 100$ | 10 |
| $100 < h \leqslant 120$ | 19 |
| $120 < h \leqslant 150$ | 9 |
| $150 < h \leqslant 200$ | 7 |

(Answers on page 131)

# 14 Probability

In this final chapter, probability is taught from the very beginning, as though the students have never seen it before.

It starts with a discussion of what probability is, and then uses words to describe how likely an event is to happen, before moving on to expressing a probability as a number between 0 and 1.

The P notation is introduced to mean 'the probability of', and simple examples using balls, coins, dice, discs, cards, spinners and sweets are presented.

The more complicated concepts of relative frequency, mutually exclusive events and independent events are then introduced.

The second half of the chapter looks at diagrams that can be used to help to solve probability problems. These are:

- sample space diagrams
- tree diagrams
- Venn diagrams.

Conditional probability is also studied.

## Prerequisite knowledge

This chapter assumes that the students are familiar with:

- two-way tables
- fractions
- basic items commonly used in probability questions, such as:
  - coins
  - dice
  - playing cards
  - spinners.

# Investigations, activities and puzzles

## Estimating pi

You can use relative frequency to estimate the value of $\pi$.

On a fairly big piece of paper, draw a quarter circle inside a square.

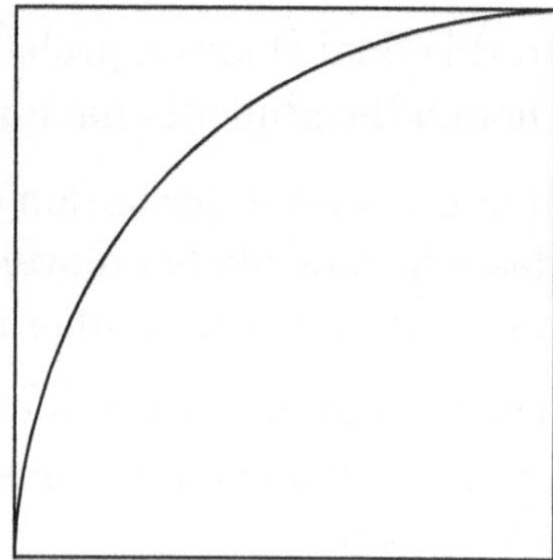

Drop drawing pins (or something else that's quite small) randomly onto the paper.

Work out an estimate of the probability that a pin will land in the quarter circle using relative frequency, where

$$\text{relative frequency} = \frac{\text{number of times pins land in quarter circle}}{\text{total number of pin drops}}$$

You know that the quarter circle has radius $r$, so the area of the square is $r^2$.

Therefore, the area of the quarter circle is $\frac{1}{4}\pi r^2$.

This means that the probability of a pin landing in the quarter circle should be $\frac{1}{4}\pi$.

Multiply your relative frequency by 4 to obtain an estimate of $\pi$.

How accurate was your estimate?

### Notes on this activity

This is called the Monte Carlo method for estimating $\pi$ and it is well documented online.

It is important for students to realise that in order for the method to work, the pins really do need to be randomly dropped onto the paper. If this method fails badly, it is probably because of a lack of randomness in the pin dropping, rather than in the method itself.

## The birthday problem

How big does a group of people need to be for the probability of two of them sharing the same birthday to be less than a half?

### Notes on this activity

It is well known among teachers of mathematics that the answer to this question is that you only need 23 people in a group for there to be more than a 50% probability of two of them having the same birthday.

Familiarity with this result, though, does not make it any less surprising or counterintuitive.

One good thing about this is that it can be tested quite easily at school since most classes have more than 23 students.

Start by asking students how many people they think you would need in a group to guarantee that at least two of them share a birthday. The answer is, of course, 367 (if you base it on there being 366 possible birthdays, including 29 February). This question acts as misdirection, as well as being a good illustration of the pigeonhole principle. When you then ask how many people you would need to bring the probability down from certain to 50%, students may expect the answer to be about half of 367. The answer of 23 will always come as a surprise to anyone who has not heard this before.

**Extension**

You can extend this problem by changing the percentage and asking students to work out how many people you would need for the probability to be your chosen value.

## Bertrand's box paradox

You have three boxes:

- box 1 contains two gold coins
- box 2 contains two silver coins
- box 3 contains one gold coin and one silver coin

If you choose a coin at random from one of the boxes and it is a gold coin, what is the probability that the other coin in the box is also gold?

### Notes on this activity

Bertrand's box paradox was first posed by the French mathematician Joseph Bertrand in 1889. It is quite commonly used when teaching probability because it is a simple example of a counterintuitive result that will make students think more deeply when they first study probability.

The answer to the question is $\frac{2}{3}$, although at first some students may jump to the conclusion that it must be $\frac{1}{2}$ reasoning that if a gold coin is chosen, there are only two options: either the other coin is gold or the other coin is silver.

The reason that this is not the correct way to think about the situation is because initially there are three gold coins that could be chosen, and two out of the three of them are in a box with another gold coin.

For this reason, this puzzle feels quite similar to the more well-known Monty Hall problem, which you could then move on to. One advantage of the Monty Hall problem is that it is now so well known that there are simulators online that enable you to run a huge number of trials very quickly to verify the result. There are also lots of videos available that explain the solution.

# Glossary

### Mutually exclusive events

Mutually exclusive events are sets of events or outcomes where one of them happening means that none of the others can happen.

### Example

The outcomes of rolling a dice are mutually exclusive, since when one number lands on the top it must mean that none of the others can.

### Probability

The probability of an outcome (or event) is a measure of how likely that outcome is. A value of O means that it is impossible, whereas 1 means that it is certain; otherwise, the value must lie between 0 and 1. A probability may be given as a common fraction, a decimal or a percentage.

### Relative frequency

The experimental probability of an outcome is the value found after an activity has been done several times. It is given by:

$$\frac{\text{number of times that named outcome(s) did happen}}{\text{number of times that the activity was done}}$$

### Examples

A drawing pin might be dropped several times and a count made of whether it landed 'point up' or 'point down'. Or a biased dice might be rolled many times to determine the experimental probability of getting a 5.

### Tree diagram

A tree diagram is drawn to find and display all possible results when several outcomes are being combined.

### Example

When a coin is tossed 3 times, all possible results can be identified and displayed by using a tree diagram:

(H = head, T = tail).

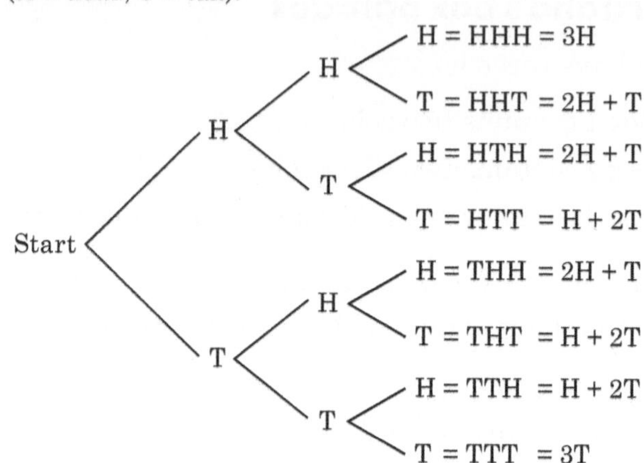

# Extension worksheet

This worksheet develops the work on combined events by introducing the binomial distribution.

When you toss two fair coins, it is easy to list the possible outcomes:

HH, HT, TH, TT.

You can draw up a simple table to show the number of ways of getting no tails, one tail and two tails:

| Number of tails | 0 | 1 | 2 |
|---|---|---|---|
| Number of ways | 1 | 2 | 1 |

You can do a similar systematic listing if you have three coins:

HHH, HHT, HTH, THH, HTT, THT, TTH, TTT

| Number of tails | 0 | 1 | 2 | 3 |
|---|---|---|---|---|
| Number of ways | 1 | 3 | 3 | 1 |

For four or more coins, this systematic approach becomes quite long and tedious.

The table for the number of ways of getting a given number of tails with four coins looks like this:

| Number of tails | 0 | 1 | 2 | 3 | 4 |
|---|---|---|---|---|---|
| Number of ways | 1 | 4 | 6 | 4 | 1 |

There is a famous mathematical diagram that shows this information for larger numbers of coins.

The first row is row 0, the next is row 1, and so on.

Each number in the diagram is the sum of the two numbers immediately above it, for example $20 = 10 + 10$.

The diagram is called *Pascal's triangle* and, in this context, can be used to determine the number of ways of obtaining a given number of heads or tails for larger numbers of throws.

```
              1
            1   1
          1   2   1
        1   3   3   1
      1   4   6   4   1
    1   5  10  10   5   1
  1   6  15  20  15   6   1
1   7  21  35  35  21   7   1
1 8 28  56  70  56  28   8  1
```

## Example 1

Imagine you want to find the number of ways of obtaining tails when you toss seven coins.

Look at row 7 in the diagram (the one that starts 1, 7). The first entry is the number of ways of tossing no tails, then one tail, two tails, and so on.

To find the *probability* of getting two tails when you toss seven coins, you multiply the appropriate number in Pascal's triangle (i.e. 21) by the probability of tossing two tails on one occasion (for example, TTHHHHH), which is:

$$21 \times 0.5^2 \times 0.5^5 = \frac{21}{128}$$

In general, this method is used in any situation where there are two distinct and mutually exclusive outcomes, where one of the outcomes is 'success' and the other is 'failure'. The probability of success in each successive *trial* must be the same. Distributions that can be modelled in this way are known as *binomial distributions*.

In the case of the coins, getting a tail is a success, and getting a head is a failure. The probability of success is constant, so you can say that P(success) = $p$ = 0.5.

**Example 2**

A fair six-sided dice is rolled eight times. Find the probability that a six is rolled on exactly three occasions.

Here, P(success) = $\frac{1}{6}$ and P(failure) = $\frac{5}{6}$

Given that there are 56 ways of getting three sixes (look at row 8 of the triangle, that starts 1 8, and go across to the fourth number), the probability can be calculated as follows:

$$\text{P(3 sixes in 8 throws)} = 56 \times \left(\frac{1}{6}\right)^3 \times \left(\frac{5}{6}\right)^5 = 0.1042 \text{ (4 d.p.)}$$

In general, you can call the number of trials in the particular model $n$ and the number of successes $r$.

The $r$th number in the $n$th row of Pascal's triangle can be calculated by using a button on your calculator called a **combination** (look for the button $nCr$ and enter $8nCr3$ to check that you get 56).

In correct notation, this combination is written as ${}^nC_r$. You can write the calculation to the above example as follows:

$$\text{P(3 sixes in 8 throws)} = {}^8C_3 \times \left(\frac{1}{6}\right)^3 \times \left(\frac{5}{6}\right)^5$$

In more general terms, you can write:

P($r$ successes in $n$ trials) = ${}^nC_r \times p^r \times q^{n-r}$, where $q$ is the probability of failure $(1 - p)$

1. Seven fair dice are rolled. Find the probability of getting each of the following:

   **a)** exactly 4 sixes      **b)** 2 or 3 sixes      **c)** 5 or more sixes

2. A coin is biased so that the probability of getting a head is 0.6. Twenty identical coins are tossed. Find the probability of getting each of the following:

   **a)** exactly 8 heads      **b)** 11, 12 or 13 heads      **c)** 17 or more heads

(Answers on page 131)

## Number

Make sure you can...

| | Revised | Secure |
|---|---|---|
| • add, subtract, multiply and divide whole numbers, decimals and fractions | | |
| • recognise: <br>   • natural numbers <br>   • integers <br>   • prime numbers <br>   • square numbers <br>   • cube numbers <br>   • common factors <br>   • common multiples <br>   • rational numbers <br>   • irrational numbers <br>   • reciprocals | | |
| • continue a given number sequence | | |
| • find the $n$th term for a given number sequence | | |
| • make estimates and round answers to an appropriate degree of accuracy | | |
| • work out upper and lower bounds for a given quantity | | |
| • use bounds in calculations | | |
| • use standard form | | |
| • expand and simplify expressions involving surds | | |
| • rationalise the denominator of a surd | | |
| • understand ratio, including map scales | | |
| • understand direct proportion, including currency conversion | | |
| • calculate percentage increase and decrease | | |
| • calculate reverse percentage change | | |
| • solve problems involving: <br>   • simple interest <br>   • compound interest | | |
| • use a calculator efficiently | | |
| • solve problems involving exponential growth and decay | | |
| • convert between fractions, decimals and percentages | | |
| • convert a recurring decimal to a fraction | | |

# Algebra and graphs

Make sure you can...

| | Revised | Secure |
|---|---|---|
| • substitute numbers into expressions and formulae | | |
| • construct and rearrange formulae | | |
| • expand brackets | | |
| • factorise expressions | | |
| • solve linear equations in one unknown | | |
| • solve practical problems using algebra | | |
| • solve a pair of linear simultaneous equations | | |
| • solve practical problems involving linear simultaneous equations | | |
| • solve simultaneous equations where one equation is nonlinear | | |
| • factorise quadratic expressions | | |
| • solve quadratic equations using:<br>  • factorisation<br>  • the quadratic formula<br>  • completing the square | | |
| • solve practical problems involving quadratic equations | | |
| • manipulate and simplify algebraic fractions | | |
| • express direct proportion using algebra | | |
| • express inverse proportion using algebra | | |
| • understand and use the rules of indices | | |
| • solve linear inequalities | | |
| • recognise, sketch and interpret graphs of these functions:<br>  • linear<br>  • quadratic<br>  • cubic<br>  • reciprocal<br>  • exponential | | |
| • plot graphs of straight lines and curves | | |
| • solve equations graphically | | |
| • plot and interpret distance–time and speed–time graphs | | |
| • calculate acceleration/deceleration and distance from travel graphs | | |
| • find the derivative of a given function | | |
| • find the gradient of a curve using differentiation | | |
| • find the coordinates of turning points on graphs | | |
| • distinguish between maximum and minimum points | | |
| • use function notation to describe simple functions | | |
| • find the inverse of a given function | | |
| • find composite functions | | |

# Coordinate geometry

Make sure you can...

| | Revised | Secure |
|---|---|---|
| • obtain the equation of a straight line in the form $y = mx + c$ | | |
| • calculate the gradient of a straight line given two coordinates | | |
| • find the length of a line segment given two coordinates | | |
| • find the midpoint of a line segment given two coordinates | | |
| • find the gradient of a parallel line | | |
| • find the equation of a parallel line | | |
| • find the gradient of a perpendicular line | | |
| • find the equation of a perpendicular line | | |
| • find the perpendicular distance from a point to a line | | |

# Geometry

Make sure you can...

| | Revised | Secure |
|---|---|---|
| • measure and draw lines and angles | | |
| • construct triangles given three sides | | |
| • find missing angles in:<br>  • triangles<br>  • quadrilaterals<br>  • irregular polygons<br>  • regular polygons | | |
| • know and use rules for angles:<br>  • at a point<br>  • on a straight line<br>  • in parallel lines | | |
| • know and use circle theorems | | |
| • recognise and describe rotational symmetry | | |
| • recognise and describe line symmetry | | |
| • identify planes of symmetry in 3D shapes | | |
| • understand and describe what is meant by similarity | | |
| • calculate missing lengths in similar shapes | | |
| • calculate missing areas in similar shapes | | |
| • calculate missing volumes in similar solids | | |
| • draw and recognise nets of solids | | |
| • use and draw three-figure bearings | | |
| • apply scale drawings to solve problems | | |

# Mensuration

Make sure you can...

| | Revised | Secure |
| --- | --- | --- |
| • calculate the perimeter of a rectangle | | |
| • calculate the area of a rectangle | | |
| • calculate the perimeter of a triangle | | |
| • calculate the area of a triangle | | |
| • calculate the perimeter of a parallelogram | | |
| • calculate the area of a parallelogram | | |
| • calculate the perimeter of a trapezium | | |
| • calculate the area of a trapezium | | |
| • calculate the area of a circle | | |
| • calculate the circumference of a circle | | |
| • calculate arc length | | |
| • calculate sector area | | |
| • calculate the surface area and volume of a:<br>  • cone<br>  • pyramid<br>  • prism<br>  • cylinder<br>  • sphere | | |
| • convert between units including units of area and volume | | |
| • calculate the area of compound shapes | | |
| • calculate the volume of compound solids | | |

# Trigonometry

Make sure you can...

| | Revised | Secure |
|---|---|---|
| • use Pythagoras' theorem | | |
| • apply the sine, cosine and tangent ratios to find unknown sides | | |
| • apply inverse sine, cosine and tangent to find unknown angles | | |
| • calculate angles of elevation and depression | | |
| • use Pythagoras' theorem and trigonometry in three dimensions | | |
| • recognise, sketch and interpret graphs of trigonometric functions | | |
| • solve simple trigonometric equations | | |
| • use exact trigonometric values in calculations | | |
| • understand and use the sine rule | | |
| • understand and use the cosine rule | | |
| • understand and use the formula for the area of a triangle | | |

# Transformations and vectors

Make sure you can...

| | Revised | Secure |
|---|---|---|
| • add and subtract vectors | | |
| • multiply a vector by a scalar | | |
| • calculate the magnitude of a vector | | |
| • interpret and show vectors geometrically | | |
| • recognise parallel vectors | | |
| • use position vectors | | |
| • apply the following transformations to a shape:<br>  • rotation<br>  • reflection<br>  • translation<br>  • enlargement | | |
| • apply two or more transformations to a shape | | |

# Probability

Make sure you can...

| | Revised | Secure |
|---|---|---|
| • describe sets using appropriate notation and language | | |
| • draw Venn diagrams | | |
| • calculate simple probabilities | | |
| • understand how to calculate the probability of an event *not* happening | | |
| • calculate relative frequency | | |
| • calculate expected frequency | | |
| • calculate the probability of combined events using:<br>  • sample space diagrams<br>  • tables<br>  • tree diagrams<br>  • Venn diagrams | | |
| • calculate conditional probabilities using:<br>  • Venn diagrams<br>  • tree diagrams<br>  • tables | | |

# Statistics

Make sure you can...

| | Revised | Secure |
|---|---|---|
| • construct the following types of chart:<br>  • bar chart<br>  • pie chart<br>  • pictogram<br>  • frequency polygon<br>  • histogram<br>  • cumulative frequency diagram<br>  • stem-and-leaf diagram | | |
| • calculate the mean, median, mode and range from discrete data | | |
| • calculate the mean of grouped continuous data | | |
| • calculate lower and upper quartiles | | |
| • calculate interquartile range | | |
| • estimate quartiles and interquartile range from a cumulative frequency diagram | | |
| • compare data sets using diagrams and calculations | | |
| • draw and interpret a scatter diagram | | |
| • understand what is meant by positive, negative and no correlation | | |
| • draw and interpret a line of best fit | | |

1. How many mm are there in 1 m 1 cm?

   **A** 1001            **B** 1110

   **C** 1010            **D** 1100

2. In standard form the value of $200 \times 80\,000$ is:

   **A** $16 \times 10^6$            **B** $1.6 \times 10^9$

   **C** $1.6 \times 10^7$            **D** $1.6 \times 10^8$

3. A person paid tax on \$9000 at 30%. They paid the tax in 12 equal payments. Each payment was:

   **A** \$2.25            **B** \$22.50

   **C** \$225            **D** \$250

4. The approximate value of $\dfrac{3.96 \times (0.5)^2}{97.1}$ is:

   **A** 0.01            **B** 0.02

   **C** 0.04            **D** 0.1

5. Given that $\dfrac{3}{n} = 5$, then $n =$

   **A** 2            **B** $-2$

   **C** $1\dfrac{2}{3}$            **D** 0.6

6. How many prime numbers are there between 30 and 40?

   **A** 0            **B** 1

   **C** 2            **D** 3

7. A man is paid \$600 per week after a pay rise of 20%. What was he paid before?

   **A** \$480            **B** \$500

   **C** \$540            **D** \$580

8. A car travels for 20 minutes at 45 km/h and then for 40 minutes at 60 km/h. The average speed for the whole journey is:

   **A** $52\dfrac{1}{2}$ km/h            **B** 50 km/h

   **C** 54 km/h            **D** 55 km/h

**9.** What is the value of the expression $(x - 2)(x + 4)$ when $x = -1$?

   **A** 9                                   **B** −9

   **C** 5                                   **D** −5

**10.** $\dfrac{1}{3} + \dfrac{2}{5} =$

   **A** $\dfrac{2}{8}$                                 **B** $\dfrac{3}{8}$

   **C** $\dfrac{3}{15}$                              **D** $\dfrac{11}{15}$

**11.** 800 decreased by 5% is:

   **A** 795                                 **B** 640

   **C** 760                                 **D** 400

**12.** Given that $a = \dfrac{3}{5}, b = \dfrac{1}{3}, c = \dfrac{1}{2}$ then:

   **A** $a < b < c$                      **B** $a < c < b$

   **C** $a > b > c$                      **D** $a > c > b$

**13.** What is the value of $x$ satisfying the simultaneous equations

   $3x + 2y = 13$

   $x - 2y = -1?$

   **A** 7                                     **B** 3

   **C** $3\dfrac{1}{2}$                               **D** 2

**14.** I start with $x$, then square it, multiply by 3 and finally subtract 4.

   The final result is:

   **A** $(3x)^2 - 4$                     **B** $(3x - 4)^2$

   **C** $3x^2 - 4$                       **D** $3(x - 4)^2$

(Answers on page 132)

# Multiple choice revision test 2

1. The circumference of a circle is $16\pi$ cm.

   The radius, in cm, of the circle is:

   **A**  2                      **B**  4

   **C**  $\dfrac{4}{\pi}$            **D**  8

2. In the triangle the value of $\cos x$ is:

   **A**  0.8               **B**  1.333

   **C**  0.75            **D**  0.6

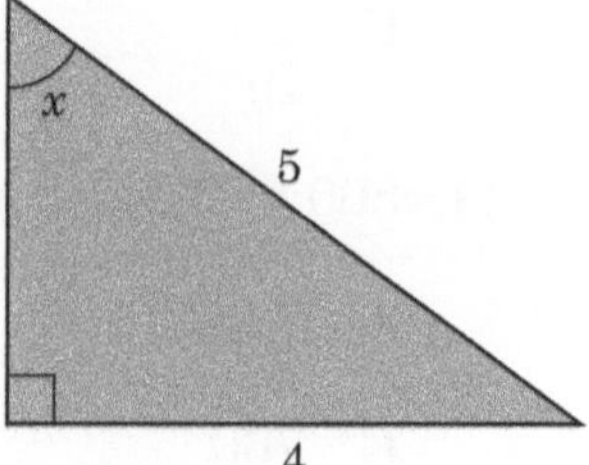


3. The solutions of the equation $(x - 3)(2x + 1) = 0$ are:

   **A**  $-3, \dfrac{1}{2}$         **B**  $3, -2$

   **C**  $3, -\dfrac{1}{2}$         **D**  $-3, -2$

4. Cube A has side length 2 cm. Cube B has side length 4 cm.

   $$\dfrac{\text{volume of B}}{\text{volume of A}} =$$

   **A**  2                      **B**  4

   **C**  8                      **D**  16

5. The equation $ax^2 + x - 6 = 0$ has a solution $x = -2$.

   What is the value of $a$?

   **A**  1                      **B**  $-2$

   **C**  $\sqrt{2}$              **D**  2

6. The shaded area, in $\text{cm}^2$, is:

   **A**  $16 - 2\pi$         **B**  $16 - 4\pi$

   **C**  $\dfrac{4}{\pi}$            **D**  $64 - 8\pi$

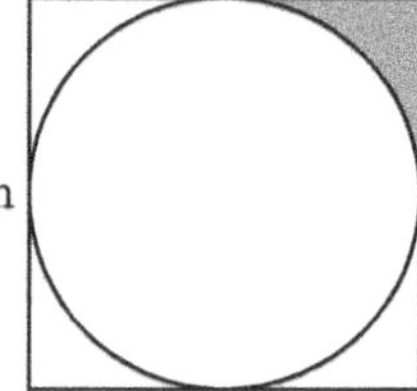


7. The perimeter of a square is 36 cm.

   What is its area?

   **A**  $36 \text{ cm}^2$         **B**  $324 \text{ cm}^2$

   **C**  $81 \text{ cm}^2$         **D**  $9 \text{ cm}^2$

8.  What is the sine of 45°?

    **A** 1

    **C** $\dfrac{1}{\sqrt{2}}$

    **B** $\dfrac{1}{2}$

    **D** $\sqrt{2}$

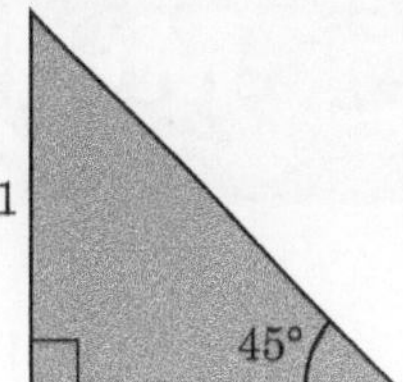


9.  The *larger* angle between south-west and east is:

    **A** 225°

    **C** 135°

    **B** 240°

    **D** 315°

10. The graph of $y = (x - 3)(x - 2)$ intersects the $y$-axis at P.

    The coordinates of P are:

    **A** (0, 6)

    **C** (2, 0)

    **B** (6, 0)

    **D** (3, 0)

11. Find the value of length $x$.

    **A** 6

    **C** $\sqrt{44}$

    **B** 5

    **D** $\sqrt{18}$

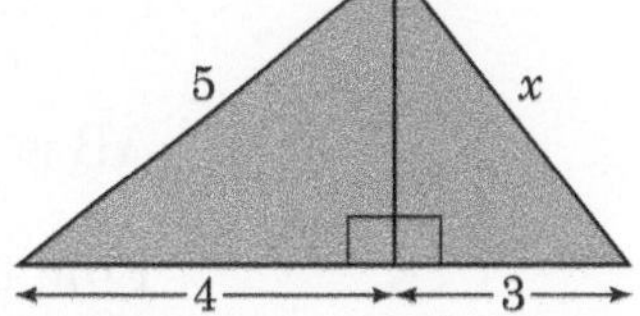


12. The bearing of A from B is 120°. What is the bearing of B from A?

    **A** 060°

    **C** 240°

    **B** 120°

    **D** 300°

13. The largest number of 1 cm cubes that will fit inside a cubical box of side 1 m is:

    **A** $10^3$

    **C** $10^8$

    **B** $10^6$

    **D** $10^{12}$

14. Which of these values of $x$ is a solution to $x + y = 8$ and $x^2 - 3x = y - 5$?

    **A** 2

    **C** 4

    **B** 3

    **D** 5

(Answers on page 132)

1. The line $y = 2x - 1$ intersects the $x$-axis at P.

   The coordinates of P are:

   **A** $(0, -1)$          **B** $\frac{1}{2}, 0$

   **C** $-\frac{1}{2}, 0$          **D** $(-1, 0)$

2. The formula $b + \dfrac{x}{a} = c$ is rearranged to make $x$ the subject.

   What is the value of $x$?

   **A** $a(c - b)$          **B** $ac - b$

   **C** $\dfrac{c - b}{a}$          **D** $ac + ab$

3. AB is a diameter of the circle.

   Find the size of angle BCO.

   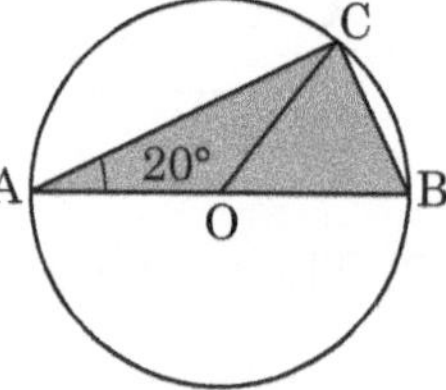
   

   **A** $70°$          **B** $20°$

   **C** $60°$          **D** $50°$

4. The sine of which of these angles is *not* 0.5?

   **A** $30°$          **B** $-30°$

   **C** $150°$          **D** $-210°$

5. The point $(3, -1)$ is reflected in the line $y = 2$.

   The new coordinates are:

   **A** $(3, 5)$          **B** $(1, -1)$

   **C** $(3, 4)$          **D** $(0, -1)$

6. What are the coordinates of the point $(1, -1)$ after reflection in the line $y = x$?

   **A** $(-1, 1)$          **B** $(1, 1)$

   **C** $(-1, -1)$          **D** $(1, -1)$

7. What kind of function is this?

   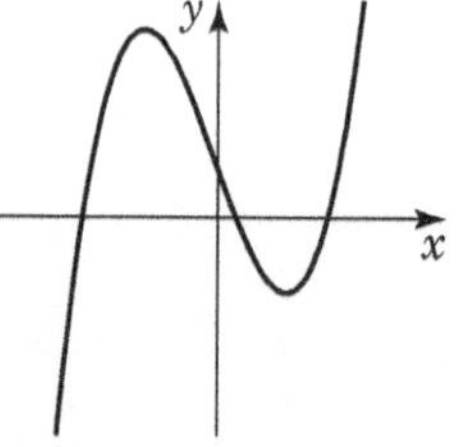
   

   **A** quadratic          **B** exponential

   **C** cubic          **D** reciprocal

**8.** What is the length of BC?

   **A** 8.91            **B** 8.93

   **C** 8.95            **D** 8.97

**9.** What values of $x$ satisfy the inequality $2 - 3x > 1$?

   **A** $x < -\dfrac{1}{3}$        **B** $x > -\dfrac{1}{3}$

   **C** $x > \dfrac{1}{3}$         **D** $x < \dfrac{1}{3}$

**10.** Given $2^x = 3$ and $2^y = 5$, the value of $2^{x+y}$ is:

   **A** 15              **B** 8

   **C** 4               **D** 125

**11.** $a = \sqrt{\dfrac{m}{x}}$

   $x =$

   **A** $a^2\, m$           **B** $a^2 - m$

   **C** $\dfrac{m}{a^2}$            **D** $\dfrac{a^2}{m}$

**12.** What is the length of BC?

   **A** 5.95            **B** 5.96

   **C** 5.97            **D** 5.98

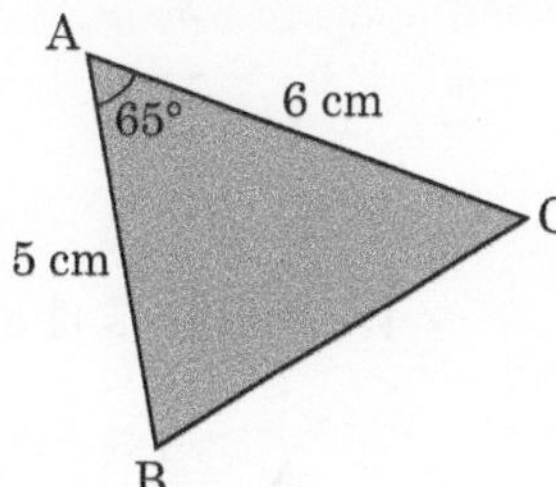


(Answers on page 132)

1. The mean mass of a group of 11 people is 70 kg. What is the mean mass of the remaining group when a person of mass 90 kg leaves?

   **A**  80 kg                               **B**  72 kg

   **C**  68 kg                               **D**  62 kg

2. Two discs are randomly taken, without replacement, from a bag containing 3 red discs and 2 blue discs. What is the probability of taking 2 red discs?

   **A**  $\dfrac{9}{25}$                         **B**  $\dfrac{1}{10}$

   **C**  $\dfrac{3}{10}$                         **D**  $\dfrac{2}{5}$

3. Four people each toss a coin. What is the probability that the fourth person will toss a tail?

   **A**  $\dfrac{1}{2}$                         **B**  $\dfrac{1}{4}$

   **C**  $\dfrac{1}{8}$                         **D**  $\dfrac{1}{16}$

4. What is the magnitude of the vector $\begin{pmatrix} 4 \\ -3 \end{pmatrix}$ ?

   **A**  –5                              **B**  1

   **C**  25                              **D**  5

5. What kind of correlation does this scatter diagram show?

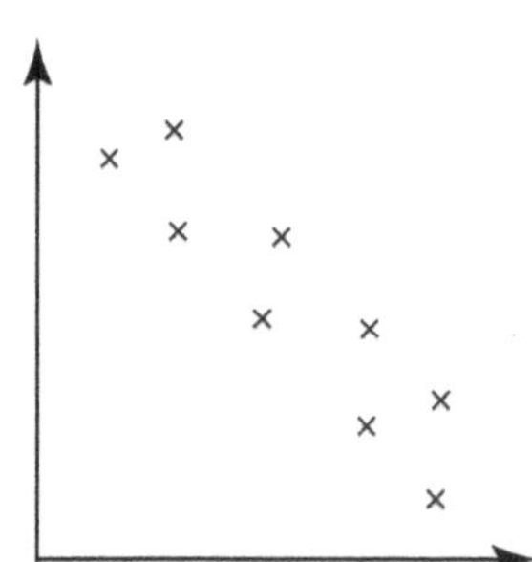

   **A**  Nonlinear correlation

   **B**  Positive correlation

   **C**  Negative correlation

   **D**  No correlation

**6.** What does the shaded area in the Venn diagram represent?

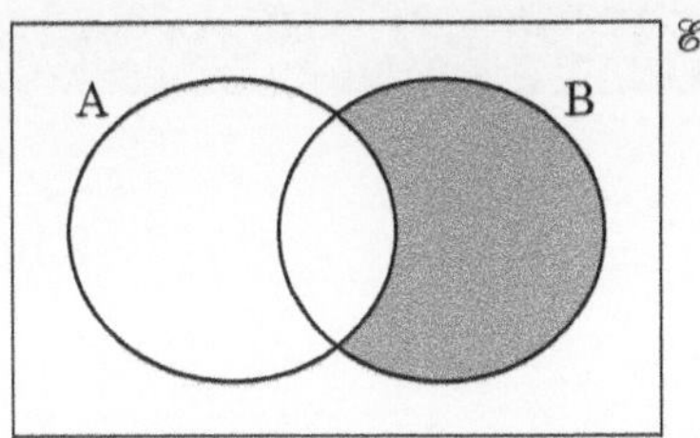

| | |
|---|---|
| **A** $A' \cup B$ | **B** $A \cap B'$ |
| **C** $A' \cap B$ | **D** $(A \cap B)'$ |

**7.** Two dice numbered 1 to 6 are thrown together and their scores are added. What is the probability that the sum will be 12?

| | |
|---|---|
| **A** $\dfrac{1}{6}$ | **B** $\dfrac{1}{12}$ |
| **C** $\dfrac{1}{18}$ | **D** $\dfrac{1}{36}$ |

**8.** Which of these vectors is not parallel to the other three?

| | |
|---|---|
| **A** $\begin{array}{c} 1 \\ 4 \end{array}$ | **B** $\begin{array}{c} 3 \\ 12 \end{array}$ |
| **C** $\begin{array}{c} -2 \\ -8 \end{array}$ | **D** $\begin{array}{c} -1 \\ 4 \end{array}$ |

**9.** The probability of an event occurring is 0.35. The probability of the event not occurring is:

| | |
|---|---|
| **A** $\dfrac{1}{0.35}$ | **B** 0.65 |
| **C** 0.35 | **D** 0 |

**10.** The mean of four numbers is 12. The mean of three of the numbers is 13. What is the fourth number?

| | |
|---|---|
| **A** 9 | **B** 12.5 |
| **C** 7 | **D** 1 |

**11.** If $f(x) = x^2 - 3$, then $f(3) - f(-1) =$

| | |
|---|---|
| **A** 5 | **B** 10 |
| **C** 8 | **D** 9 |

**12.** A pie chart is drawn to show the different colours of cars in a car park. If the blue cars take up 40° in the chart and there are 37 blue cars, how many cars are there in total?

| | |
|---|---|
| **A** 370 | **B** 93 |
| **C** 333 | **D** 1480 |

(Answers on page 132)

# Examination-style Paper 2 (Non-calculator)

## Paper 2 Non-calculator (Extended)

2 hours

- You must answer **all** questions.
- Do **not** use a calculator.
- You will need geometrical instruments.
- All necessary working should be shown clearly.
- The total mark for this paper is 100.

### List of formulae

| | |
|---|---|
| Area, $A$, of circle of radius $r$. | $A = \pi r^2$ |
| Circumference, $C$, of circle of radius $r$. | $C = 2\pi r$ |
| Curved surface area, $A$, of cylinder of radius $r$, height $h$. | $A = 2\pi r h$ |
| Volume, $V$, of cylinder of radius $r$, height $h$. | $V = \pi r^2 h$ |
| Curved surface area, $A$, of cone of radius $r$, sloping edge $l$. | $A = \pi r l$ |
| Volume, $V$, of cone of radius $r$, height $h$. | $V = \dfrac{1}{3}\pi r^2 h$ |
| Surface area, $A$, of sphere of radius $r$. | $A = 4\pi r^2$ |
| Volume, $V$, of sphere of radius $r$. | $V = \dfrac{4}{3}\pi r^3$ |
| Volume, $V$, of prism, cross-sectional area $A$, length $l$. | $V = Al$ |
| Volume, $V$, of pyramid, base area $A$, height $h$. | $V = \dfrac{1}{3}Ah$ |
| For the equation $\quad ax^2 + bx + c = 0$ where $a \neq 0$ | $x = \dfrac{-b \pm \sqrt{b^2 - 4ac}}{2a}$ |
| Area, $A$, of triangle, base $b$, height $h$. | $A = \dfrac{1}{2}bh$ |

For the triangle shown,

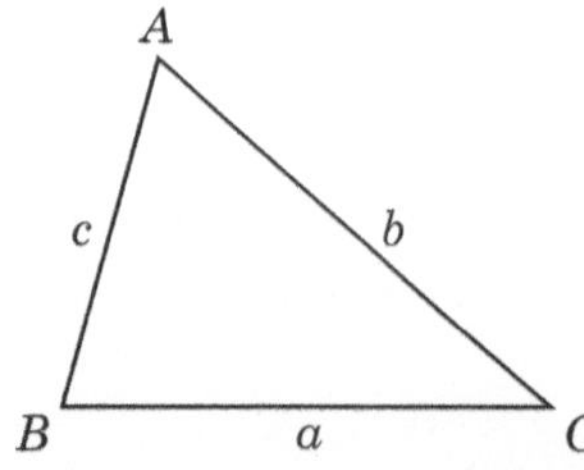

$$\frac{a}{\sin A} = \frac{b}{\sin B} = \frac{c}{\sin C}$$

$$a^2 = b^2 + c^2 - 2bc\cos A$$

$$\text{Area} = \frac{1}{2}ab\sin C$$

Do **not** use a calculator.

1.  Write 76.4996 correct to 4 significant figures.

_________ m **[1]**

2.  **a)**  Write down the order of rotational symmetry of the shape below.

_________ m **[1]**

  **b)**  Draw the lines of symmetry on the shape.                         **[1]**

3.  Given that there are 22 elements in total, complete the Venn diagram.        **[3]**

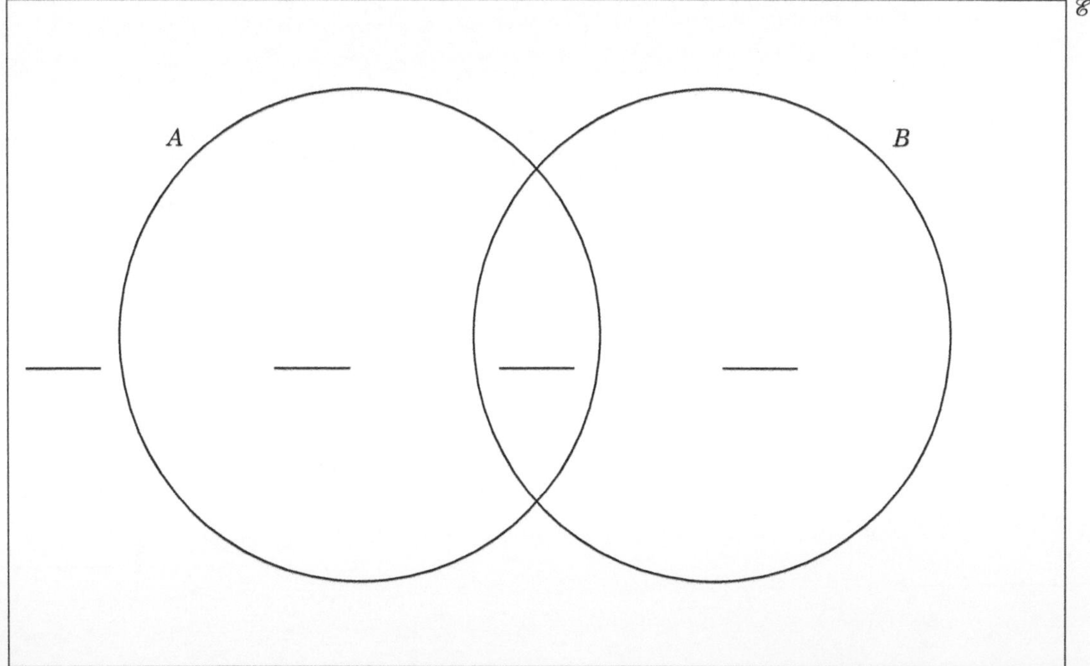


n($A \cup B$) = 18, n($A \cap B'$) = 7 and n($A$) = 11

4. A triangular prism has length 12 cm.

   The cross-section is an equilateral triangle of side length 6 cm.

   Find the exact value of the surface area of the prism.

   __________ [3]

5. Aliyah invests $700 for 7 months at 5% simple interest per year.

   Calculate the interest she receives.

   $ __________ [2]

6. Anna changes 380 euros into dollars.

   The exchange rate is 1 euro = $1.10

   Calculate the number of dollars Anna receives.

   $ __________ [1]

7. $OABC$ is a rhombus.

   $\overrightarrow{OA} = \mathbf{a}$ and $\overrightarrow{OC} = \mathbf{c}$.

   $M$ is the midpoint of $OB$. Find $\overrightarrow{MC}$ in terms of $\mathbf{a}$ and $\mathbf{c}$.

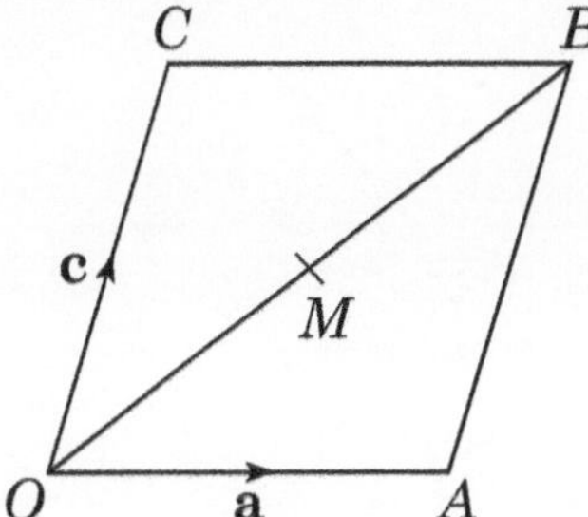


_________ [2]

8. Solve the inequality $3 - 5x > 2(3x + 5)$.

_________ [3]

9. The distance of Uranus from the sun is $2\,871\,000\,000$ km.

   Write the distance in standard form.

_________ km [1]

**10.**

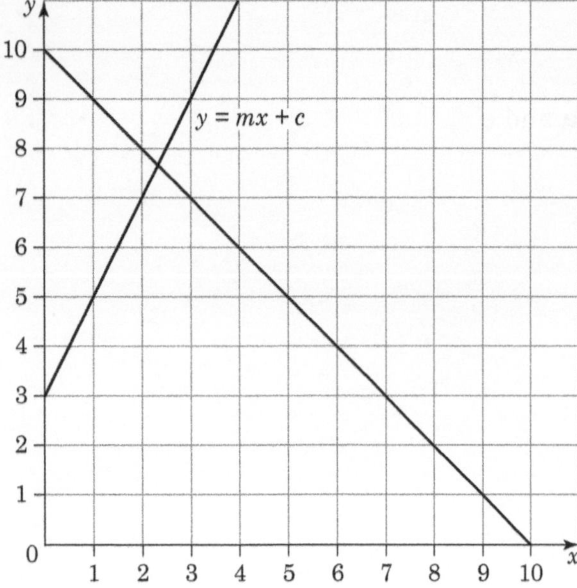


**a)** One of the lines in the diagram is labelled $y = mx + c$.

Find the values of $m$ and $c$.

$m =$ _________ **[1]**

$c =$ _________ **[1]**

**b)** A region of the diagram is defined by these four inequalities:

$$y \leqslant mx + c \qquad x + y \leqslant 10 \qquad x \geqslant 2 \quad y \geqslant 1$$

Show the defined region by shading all the unwanted regions of the diagram. Label the defined region with the letter **R**. **[2]**

**11.** Convert 0.23 m² into cm².

_________ cm² **[1]**

**12.** Work out $3\dfrac{2}{5} + 4\dfrac{1}{2}$

Give your answer as a mixed number in its simplest form.

_________ **[3]**

**13.** A train departs Paris at 1030 and arrives in Brussels at 1325.

How long does the journey take?

__________ [1]

**14.** For each of the following sequences, write down the next term.

**a)** 1, 2, 3, 5, 8, ...  __________ [1]

**b)** $x^5$, $5x^4$, $10x^3$, $15x^2$, ...  __________ [1]

**c)** 1, 4, 16, 64, ...  __________ [1]

**15. a)** Simplify $\sqrt{12} + \sqrt{108}$

__________ [2]

**b)** Rationalise the denominator.

$$\frac{1}{\sqrt{3} - 1}$$

__________ [2]

**16.** The quantity $y$ varies as the square of $(x + 1)$.

$y = 16$ when $x = 0$. Find $y$ when $x = 2$.

 [3]

**17.** Solve the simultaneous equations

$3x - 2y = 16$

$2x + 3y = 2$

_________ [3]

**18.** Simplify $\left(9x^2\right)^{\frac{3}{2}}$

_________ [2]

**19. a)** Factorise fully $cx^3 - dx^3$.

_________ [1]

**b)** Make $x$ the subject of the formula

$cx^3 - dx^3 - a = b^2$.

_________ [2]

**20.**

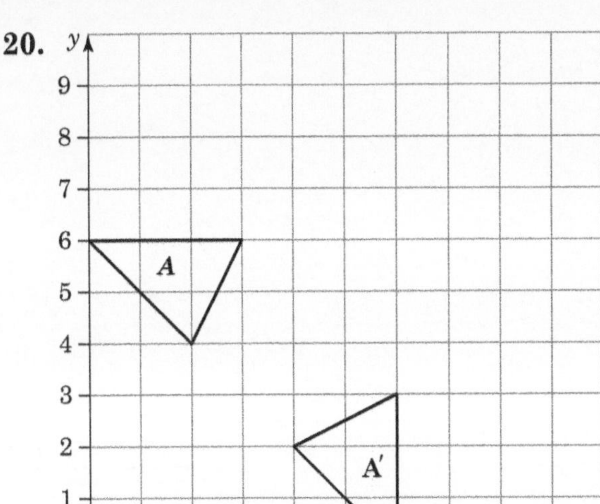

**a)** Describe fully the single transformation that maps triangle $A$ onto triangle $A'$.

_______________________________________________________

_______________________________________________________

[2]

**b)** On the same grid, draw the image of triangle $A'$ after a rotation by 90° anticlockwise about the point (5, 4). Label the image $A''$. [2]

**21.** $f(x) = 3x^3 - 4x^2 + 2x - 1.$

    **a)** Find $f'(x)$.

                                            __________ **[3]**

    **b)** Hence write down the gradient of the curve $y = f(x)$ at the point where $x = 2$.

                                            __________ **[1]**

**22.** The table gives the values of $f(x) = 3^x$ for $-2 \leqslant x \leqslant 4$.

| $x$ | $-2$ | $-1$ | 0 | 1 | 2 | 3 | 4 |
|---|---|---|---|---|---|---|---|
| $f(x)$ | $a$ | $\dfrac{1}{3}$ | $b$ | 3 | 9 | $c$ | 81 |

    **a)** Find the values of $a$, $b$ and $c$.

                                     $a =$ __________ **[1]**

                                     $b =$ __________ **[1]**

                                     $c =$ __________ **[1]**

**b)** On the grid below, draw the graph of $y = f(x)$ for $-2 \leqslant x \leqslant 4$.

Use a scale of 1 cm to 1 unit on the $x$-axis and 1 cm to 5 units on the $y$-axis.

[5]

**c)** Use your graph to solve the equation $3^x = 20$.

$x =$ _________ [1]

**d)** What value does $f(x)$ approach as $x$ decreases?

_________ [1]

**e)** By drawing a tangent, estimate the gradient of the graph of $y = f(x)$ when $x = 1.2$.

_________ [3]

**f)** On the same graph as in part (b), draw the graph of $y = 3x + 2$ for $0 \leqslant x \leqslant 4$.

[2]

**g)** Use your graph to find the positive solution of $3^x = 3x + 2$.

_________ [2]

**23.** A box of chocolates contains some caramel and some fruit-filled chocolates.

    **a)** There are $c$ caramel chocolates, which have a total mass of 105 g.

       Write down, in terms of $c$, the mean mass of one caramel chocolate.

___________ [1]

    **b)** There are $c + 4$ fruit-filled chocolates, which also have a total mass of 105 g.

       Write down, in terms of $c$, the mean mass of one fruit-filled chocolate.

___________ [1]

    **c)** The difference between the mean mass of a caramel chocolate and the mean mass of a fruit-filled chocolate is 0.8 g.

       Show that $c$ satisfies the equation $c^2 + 4c - 525 = 0$.

[4]

    **d)** Solve the equation in part (c) to find the value of $c$.

___________ [3]

    **e)** Find the mean mass of a chocolate in the box.

___________ [3]

**24. a)**

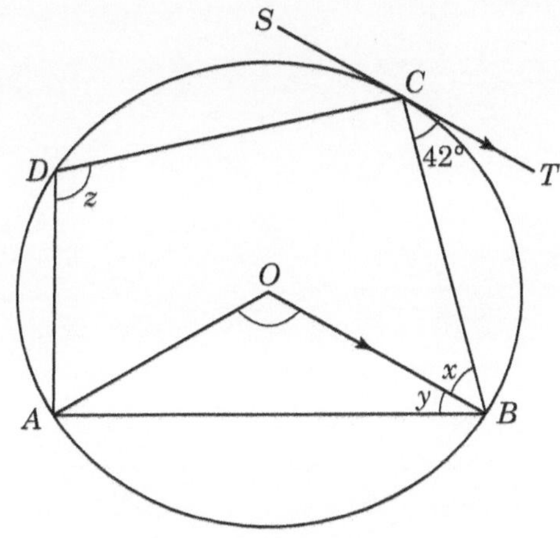

NOT TO SCALE

$A$, $B$, $C$ and $D$ lie on the edge of a circle, centre $O$.
$SCT$ is a tangent at $C$ and is parallel to $OB$.

Angle $AOB = 140°$ and angle $BCT = 42°$.

  **i)** What type of quadrilateral is $ABCD$?

_______________________________________________ **[1]**

  **ii)** Find the values of $x$, $y$ and $z$.

_______________________________________________

_______________________________________________

_______________________________________________

$x =$ _________

$y =$ _________

$z =$ _________ **[3]**

  **iii)** Write down the size of angle $OCT$.

_________ **[1]**

  **iv)** Find the size of the reflex angle $AOC$.

_______________________________________________

_________ **[1]**

**b)**

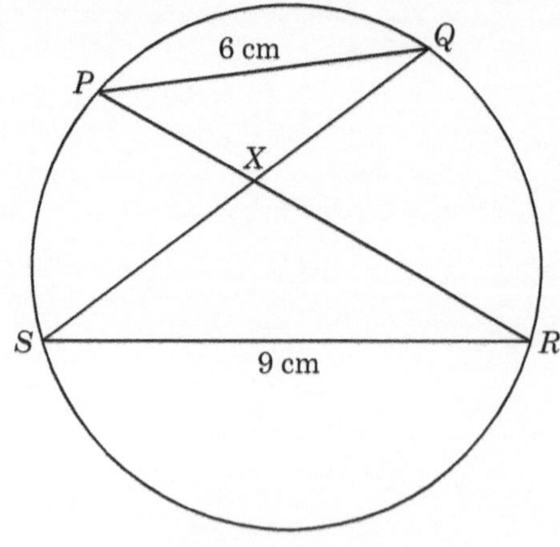


$P$, $Q$, $R$ and $S$ lie on a circle. $PQ = 6$ cm and $SR = 9$ cm. $PR$ and $SQ$ intersect at $X$.

The area of the triangle $SRX = 15$ cm$^2$.

**i)** Explain why triangles $PQX$ and $SRX$ are similar.

[3]

**ii)** Calculate the area of triangle $PQX$.

_________ [2]

**25.** A survey of 120 boys and girls was carried out and their favourite sport recorded.

The results are shown in the table.

|  | Football | Tennis | Hockey | Total |
|---|---|---|---|---|
| **Boys** | 13 | 18 | 12 | 43 |
| **Girls** | 17 | $x$ | 40 | 77 |
| **Total** | 30 | $y$ | 52 | 120 |

**a)** Write down the values of $x$ and $y$.

$x =$ _________

$y =$ _________ [1]

**b)** Find the probability that a child chosen at random likes

   **i)** football

_________________________________________________

_________ [1]

   **ii)** hockey, given that they are a girl.

_________________________________________________

_________________________________________________

_________ [2]

**c)** Find the probability that a child chosen at random is

   **i)** a boy

_________________________________________________

_________ [1]

   **ii)** a girl, given that they like football.

_________________________________________________

_________________________________________________

_________ [2]

## Paper 4 Calculator (Extended)

2 hours

- You must answer **all** questions.
- You may use a calculator.
- You will need geometrical instruments.
- All necessary working should be shown clearly.
- Unless otherwise stated, give non-exact answers to 3 significant figures. For angles in degrees, give non-exact answers to 1 decimal place.
- The total mark for this paper is 100.

### List of formulae

Area, $A$, of circle of radius $r$.

$$A = \pi r^2$$

Circumference, $C$, of circle of radius $r$.

$$C = 2\pi r$$

Curved surface area, $A$, of cylinder of radius $r$, height $h$.

$$A = 2\pi r h$$

Volume, $V$, of cylinder of radius $r$, height $h$.

$$V = \pi r^2 h$$

Curved surface area, $A$, of cone of radius $r$, sloping edge $l$.

$$A = \pi r l$$

Volume, $V$, of cone of radius $r$, height $h$.

$$V = \frac{1}{3}\pi r^2 h$$

Surface area, $A$, of sphere of radius $r$.

$$A = 4\pi r^2$$

Volume, $V$, of sphere of radius $r$.

$$V = \frac{4}{3}\pi r^3$$

Volume, $V$, of prism, cross-sectional area $A$, length $l$.

$$V = Al$$

Volume, $V$, of pyramid, base area $A$, height $h$.

$$V = \frac{1}{3}Ah$$

For the equation $ax^2 + bx + c = 0$ where $a \neq 0$

$$x = \frac{-b \pm \sqrt{b^2 - 4ac}}{2a}$$

Area, $A$, of triangle, base $b$, height $h$.

$$A = \frac{1}{2}bh$$

For the triangle shown,

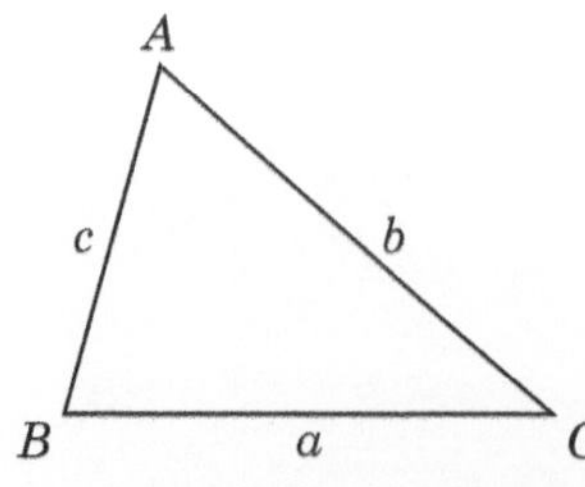

$$\frac{a}{\sin A} = \frac{b}{\sin B} = \frac{c}{\sin C}$$

$$a^2 = b^2 + c^2 - 2bc\cos A$$

$$\text{Area} = \frac{1}{2}ab\sin C$$

1. Calculate the value of $(\cos 50°)^2 + (\sin 50°)^2$.

   _______________________________________________

   _______________________________________________

   _______________________________________________

   __________ [1]

2. $f(x) = 8x$.

   **a)** Calculate f(0.4).

   _______________________________________________

   _______________________________________________

   _______________________________________________

   __________ [1]

   **b)** Write down the value of $f^{-1}(1)$.

   _______________________________________________

   __________ [1]

3. **a)**

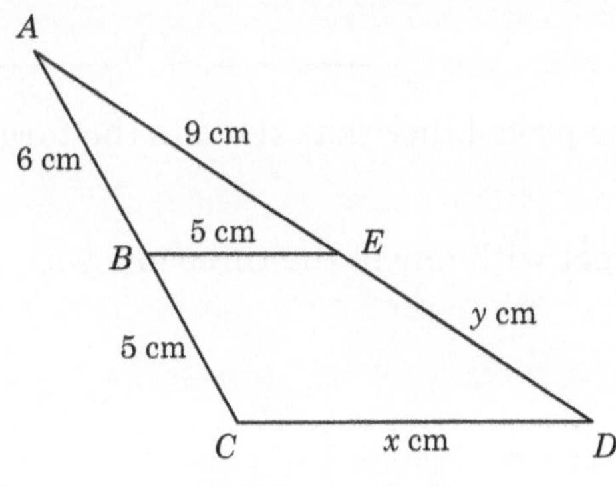


   NOT TO SCALE

   In the diagram, triangles *ABE* and *ACD* are similar.

   *BE* is parallel to *CD*.

   $AB = 6$ cm, $BC = 5$ cm, $BE = 5$ cm, $AE = 9$ cm, $CD = x$ cm and $ED = y$ cm.

   Work out the values of $x$ and $y$.

   _______________________________________________

   _______________________________________________

   _______________________________________________

   $x =$ __________ cm

   $y =$ __________ cm [4]

**b)** A spherical balloon of radius 4 cm has a surface area of $64\pi$ cm². It is further inflated until its radius is 16 cm.

Calculate its new surface area, leaving your answer in terms of $\pi$.

_________ cm² [2]

4. Indira works 7 hours a day, correct to the nearest 10 minutes. She works five days a week for a total of $W$ hours.

   Between what limits does $W$ lie?

_________ $\leq W$ _________ [2]

5. Katherine shoots an arrow at a target three times. The probability that she hits the target is 0.3.

   Calculate the probability that she does not hit the target with any of the three arrows.

_________ [2]

6. $\dfrac{4a}{5} - \dfrac{7a}{30} = \dfrac{5}{8}$

   Find the value of $a$.

$a =$ _________ [2]

7. Jemima played an online game that you can either win or lose. She played 400 times and won 240 of these games. She then played $x$ games and lost none.

   She won 80% of all the games she played. Find the value of $x$.

   _________ [4]

8. In a laboratory procedure, the number of microorganisms, $M$, after $n$ days is given by the formula

   $M = 500 \times 1.3^n$

   a) Complete the table of values.

   | $n$ | 0 | 1 | 2 | 3 | 4 |
   |---|---|---|---|---|---|
   | $M$ | | | | | |

   [2]

   b) After how many days does the number of microorganisms become greater than 2000?

   _________ days [1]

9. Solve the following pair of simultaneous equations, giving your answers to two decimal places.

$$y = 10 - 2x$$
$$y = 3x^2 + x + 4$$

_________ and _________ [5]

**10.** The time it takes 50 students to solve a puzzle is shown in the frequency table.

| Time, $t$ minutes | $4.0 < t \leqslant 5.5$ | $5.5 < t \leqslant 6.0$ | $6.0 < t \leqslant 7.0$ |
|---|---|---|---|
| Frequency | 15 | 18 | 17 |

**a)** Calculate an estimate of the mean time taken.

_________ **[3]**

**b)** Complete an accurate histogram to show this information.

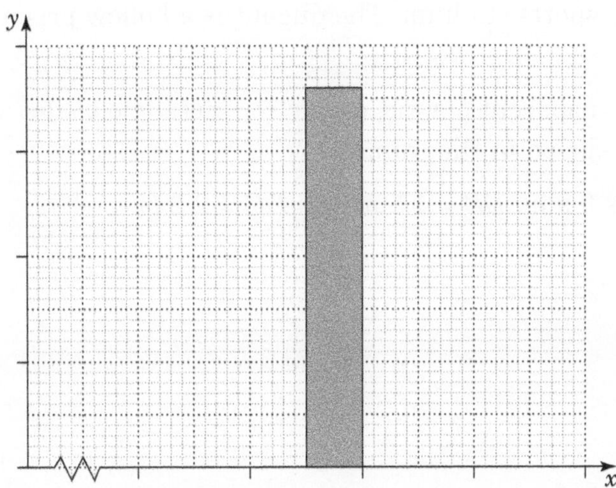

**[3]**

**c)** Find the number of students represented by $1\,\text{cm}^2$ on the histogram.

_________ **[1]**

**11.**

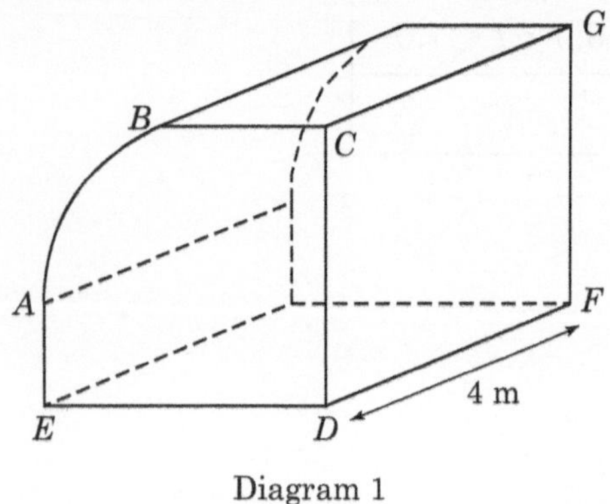

Diagram 1          Diagram 2

NOT TO SCALE

Diagram 1 shows a coaches' dugout from a sports stadium. The dugout is a hollow prism of length 4 m.

The dugout is constructed of toughened transparent plastic, except for rectangle *CDFG* which is open. The cross-section of the dugout is shown in Diagram 2.

*AB* is an arc from a circle, centre *O*, radius 0.9 m. *ED* = 1.5 m and *DC* = 2.2 m.

Calculate

**a)** the perimeter of the cross-section

_________ m **[3]**

**b)** the area of the cross-section

_________ m² **[3]**

**c)** the volume of the dugout

_________ m³ **[1]**

**b)** the total amount of plastic sheeting needed to construct the dugout.

_________ m² **[4]**

**12.** $A$ is the point $(3, 5)$ and $B$ is the point $(-2, 7)$

    **a)** Write down the midpoint of $AB$.

_________ **[1]**

    **b)** Calculate the length of $AB$.

_________ **[2]**

    **c)** Find the equation of the line that is perpendicular to $AB$ and that passes through the point $(9, 1)$.

        Give your answer in the form $y = mx + c$.

_________ **[4]**

**13. a)** Jaipreet invests $2000 at a rate of 3.5% per year **simple** interest.

Marco invests $2000 at a rate of 3% per year **compound** interest.

Work out who has the most money after 5 years, and by how much.

_________, by $ _________ **[6]**

**b)** Janice buys a car for $8500.

The value of the car depreciates at a rate of 12% per year.

Find the value of Janice's car after 4 years.

$ _________ **[3]**

**14.**

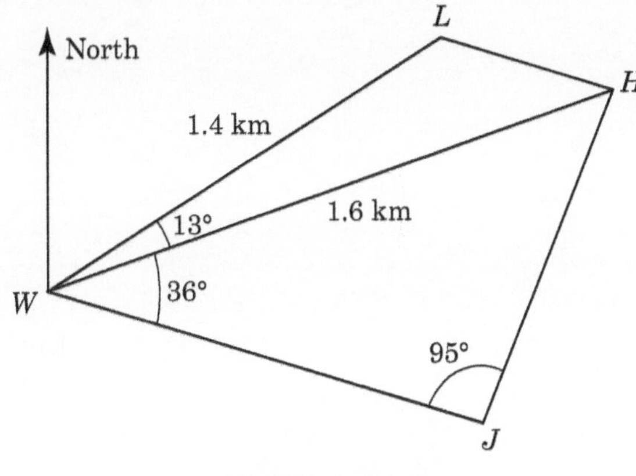


The diagram shows the positions of four orienteering markers on a course. The markers are located at the four corners of a large field.

$WL = 1.4$ km and $WH = 1.6$ km.

Angle $LWH = 13°$, angle $HWJ = 36°$ and angle $WJH = 95°$.

**a)** Calculate the distance between markers $L$ and $H$.

_________ km **[4]**

**b)** Calculate the distance between markers $W$ and $J$.

_________ km **[4]**

**c)** Calculate the area of the field *WJHL*.

_________ km² **[3]**

**d)** The bearing of marker *L* from marker *W* is 060°.

Calculate the bearing of

**i)** marker *H* from marker *W*

_________ **[1]**

**ii)** marker *W* from marker *J*.

_________ **[1]**

**e)** On a scaled map, the distance between markers *W* and *H* is 4 cm.

Calculate the scale of the map, giving your answer in the form $1:n$.

_________ **[2]**

**15.** Belinda goes to the supermarket to buy bread, juice and popcorn.

    **a) i)** The costs of the bread, juice and popcorn are in the ratio $3:4:2$.
        The cost of the bread is $3.30.

        Calculate Belinda's total bill.

$ _______ **[2]**

    **ii)** She pays with a $50 note.

        What percentage of the $50 has she got left?

_______ **[2]**

    **b)** The masses of the bread, juice and popcorn are in the ratio $2:15:4$.
    The total mass is 14.7 kg.

    Calculate the mass of the popcorn.

_______ kg **[2]**

    **c)** Calculate the mass per $ of the juice.

_______ **[3]**

    **d)** The bread cost 20% more than in the previous week.

    Calculate the cost of the bread in the previous week.

$ _______ **[2]**

**16. a)** Solve the equation $\dfrac{x-2}{3} + \dfrac{x+3}{2} = -5$

__________ [4]

**b) i)** $p = \dfrac{4}{y-1} - \dfrac{3}{y+5}$

Find the value of $p$ when $y = 2$.

__________ [1]

**ii)** Write $\dfrac{4}{y-1} - \dfrac{3}{y+5}$ as a simplified single fraction.

__________ [2]

**iii)** solve the equation $\dfrac{4}{y-1} - \dfrac{3}{y+5} = \dfrac{2}{y}$

Give your answers to two decimal places.

__________ [3]

**c)** $a = \dfrac{c}{b+1}$

Find the value of $b$ in terms of $a$ and $c$.

$b = $ __________ [3]

# Answers and mark schemes

## Chapter 1 Extension worksheet

1. 55, 5050, 500 500

2. **a)** 26    **b)** 25    **c)** 55

   **d)** 9    **e)** 7.6    **f)** 49.5

4. **a)** 338    **b)** 169    **c)** 2550

   **d)** 15 150    **e)** 40.3

## Chapter 2 Extension worksheet

1. $(-2, 0)$ and $(0, 2)$

2. $(4, 1)$ and $(1, -2)$

3. $(1, 1)$ and $(4, -2)$

4. $(-0.54, -0.09)$ and $(2.94, 6.89)$

5. $(1.35, 0.29)$ and $(-2.95, 8.91)$

6. $(-1.18, 4.18)$ and $(3.18, -0.18)$

## Chapter 3 Extension worksheet

1. $0.27 \text{ m/s}^2$ (2 s.f.)

2. 90 m

4. $s = ut + \dfrac{1}{2}at^2$;  $a, s, t, u$

5. $t = \dfrac{2s}{u+v}$

6. $v^2 = u^2 + 2as$;  $a, s, u, v$

7. $s = vt - \dfrac{1}{2}at^2$

## Chapter 4 Extension worksheet

All answers given to 3 s.f.

1. 7.81 cm

2. 7.87 cm

3. 12.9 cm

4. 29.9 cm

5. 10.1 cm

6. $d^2 - b^2 - c^2 = a^2$ (or equivalent)

7. 15.5 cm

## Chapter 5 Extension worksheet

**1.** **a)** $\frac{\neq}{2}$ **b)** $\frac{\neq}{3}$ **c)** $\frac{2\neq}{3}$ **d)** $\frac{\neq}{4}$

**e)** $\frac{3\neq}{2}$ **f)** $\frac{3\neq}{4}$ **g)** $\frac{\neq}{6}$ **h)** $\frac{\neq}{8}$

**2.** **a)** **i)** $2\pi$ cm **ii)** $4\pi$ cm$^2$

**b)** **i)** $\pi$ cm **ii)** $3\pi$ cm$^2$

**c)** **i)** $3.5\pi$ cm **ii)** $\frac{147\neq}{8}$ cm$^2$

**d)** **i)** 6.4 cm **ii)** 10.24 cm$^2$

**e)** **i)** 5.705 cm **ii)** 4.65 cm$^2$ (3 s.f.)

**3.** **a)** 2 cm **b)** 4 cm **c)** 6.37 cm

**d)** 2.15 cm **e)** 29.8 cm

**4.** **a)** 6 cm$^2$ **b)** 16 cm$^2$ **c)** 15.9 cm$^2$

**d)** 4.83 cm$^2$ (4.84 if using the rounded value of $r$ from Question **3**)

**e)** 116 cm$^2$

## Chapter 6 Extension worksheet

**1.** $(x + 3)(x - 2)(x + 5)$

**2.** $(x - 4)(x + 2)(x - 1)$

**3.** $(x + 1)(x + 7)(x - 3)$

**4.** $(x - 5)(x + 4)(x - 2)$

**5.** $(x + 2)(x - 3)(x + 6)$

**6.** $(x + 6)(x - 1)(x + 4)$

**7.** $(x - 2)(x + 8)(x - 5)$

**8.** $(x + 4)(x - 3)(x + 1)$

## Chapter 7 Extension worksheet

**1.** Three more squares are required to make the pattern have rotational symmetry of order 2.

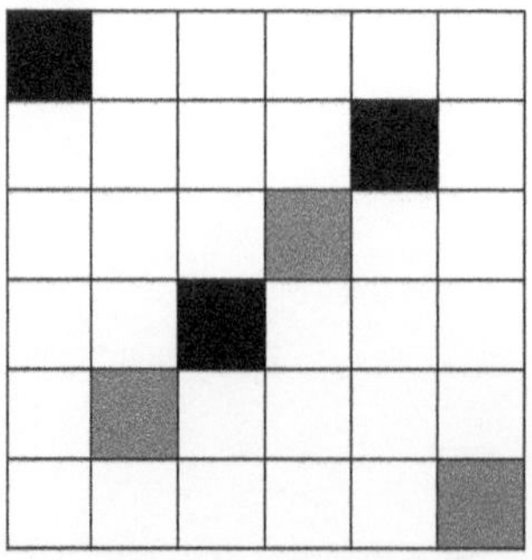

9 more squares are required for it to have rotational symmetry of order 4.

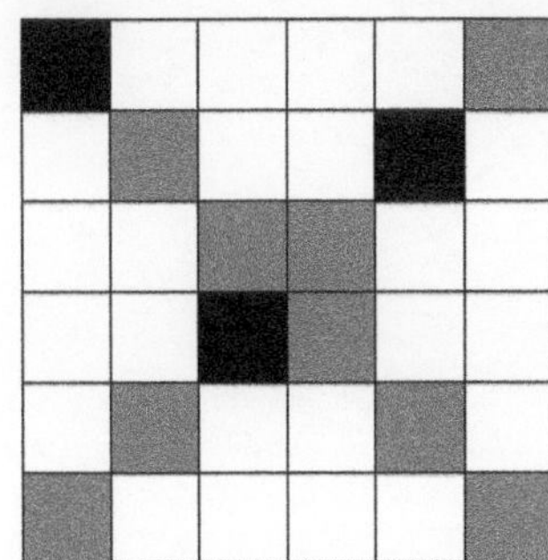

The maximum number is a total of 34, for example:

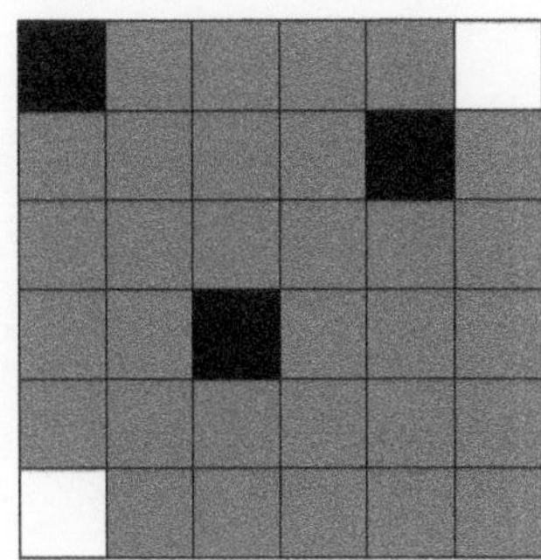

2.  Students' own answers.

## Chapter 8 Extension worksheet

1.  $a^7$

2.  $a^9$

3.  $a^6$

4.  $x^{14}$

5.  $x^1 = x$

6.  $x$

7.  square root

8.  $x^1 = x$

9.  $x$

10. cube root

11. $\dfrac{1}{4}$

12. $\dfrac{1}{5}$

13. tenth root

14. cube root squared

15. raise to the power of $a$

16. a) $x^{\frac{4}{3}}$    b) $x^{\frac{3}{2}}$    c) $x^{\frac{3}{5}}$    d) $x^{\frac{5}{6}}$    e) $x^{\frac{91}{60}}$

## Chapter 9 Extension worksheet

1. 0.5, 1, 8, 16

2. 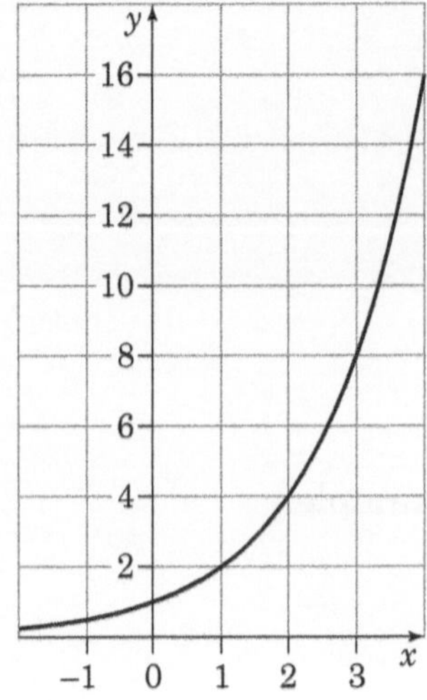

3. **a)** 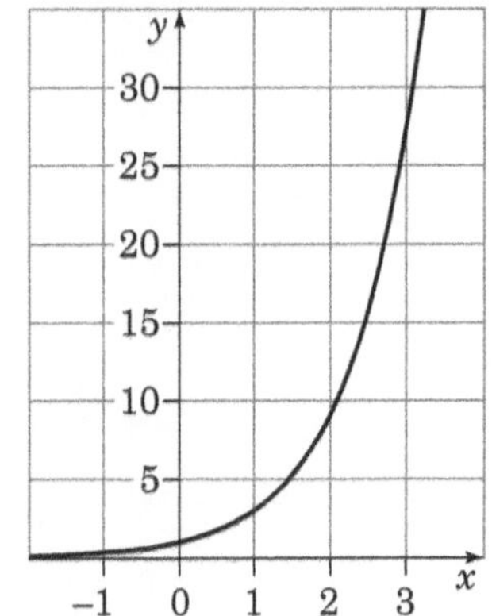   **b)** 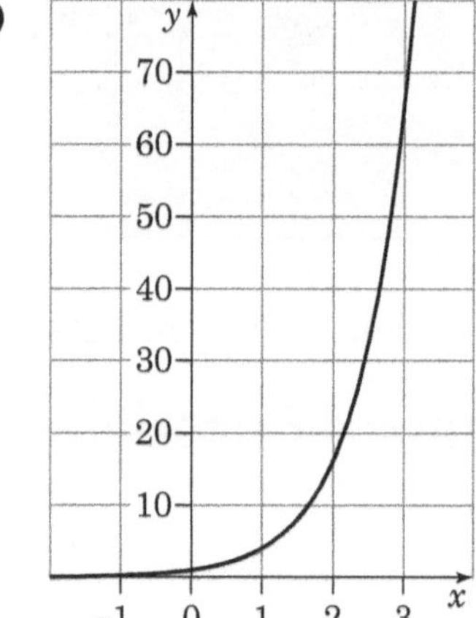

   **c)** 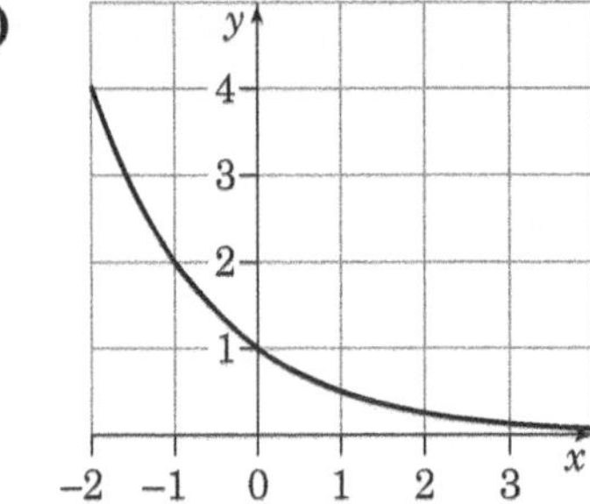

4. $y = 5$

5. **a)** $x = 1.46$   **b)** $x = 2.81$   **c)** $x = 2.18$   **d)** $x = 2.16$

   **e)** $x = 1.81$   **f)** $x = 2.32$   **g)** $x = 3.32$   **h)** $x = 1.80$

6. $y = x + 2$

7. **a)** $x = 2.45$ and $x = -2.86$   **b)** $x = 0.42$   **c)** $x = 1.39$

   **d)** $x = -1.39$ and $x = 1.69$   **e)** $x = 1.08$

   **f)** $x = 2$, $x = 4$ and $x = -0.77$

8. 0.135, 0.368, 7.39, 20.1, 54.6

**9.**

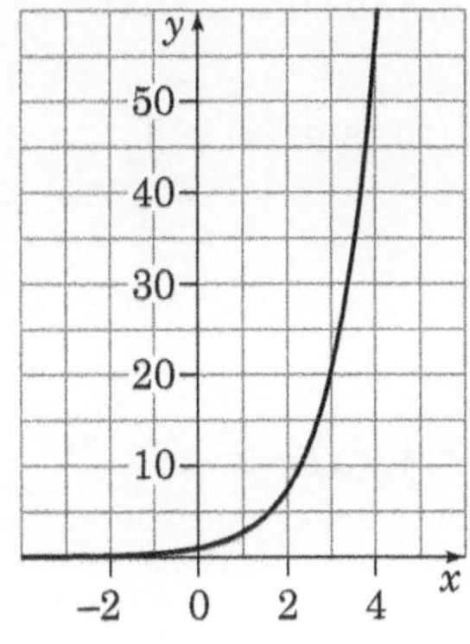

**10.** $x = 0.69$

## Chapter 10 Extension worksheet

**3.** $\sec x = \dfrac{c}{b}, \quad \operatorname{cosec} x = \dfrac{c}{a}$

**4.** $\cot x = \dfrac{b}{a}$

**6.**

| $x$ | 0° | 15° | 30° | 45° | 60° | 75° | 90° | 105° | 120° | 135° | 150° | 165° | 180° |
|---|---|---|---|---|---|---|---|---|---|---|---|---|---|
| $\sec x$ | 1 | 1.04 | 1.15 | 1.41 | 2 | 3.86 | | −3.86 | −2 | −1.41 | −1.15 | −1.04 | −1 |
| $\operatorname{cosec} x$ | | 3.86 | 2 | 1.41 | 1.15 | 1.04 | 1 | 1.04 | 1.15 | 1.41 | 2 | 3.86 | |
| $\cot x$ | | 3.73 | 1.73 | 1 | 0.58 | 0.27 | 0 | −0.27 | −0.58 | −1 | −1.73 | −3.73 | |

**7.**

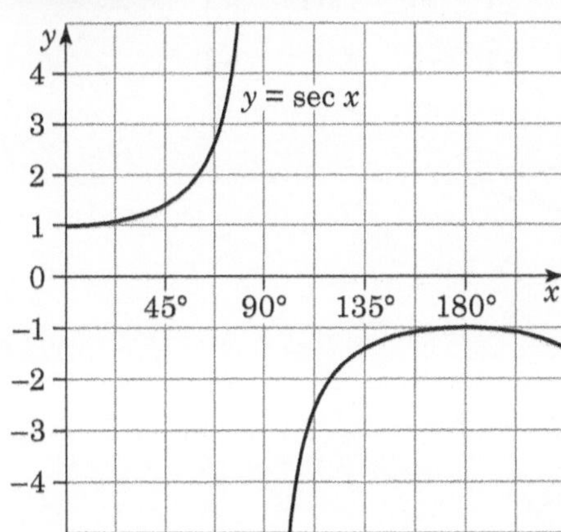
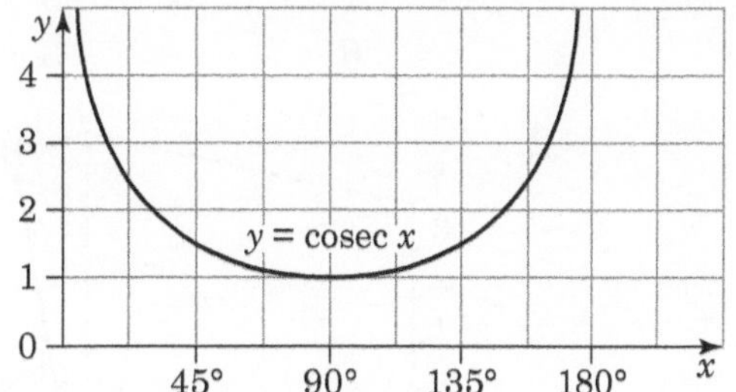
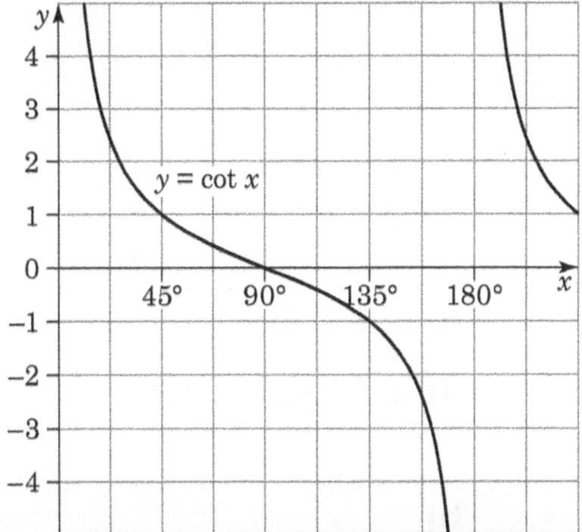

**8.** $x = 90°$; $x = 0°$ and $x = 180°$; $x = 0°$ and $x = 180°$

**9. a)** $x = 26.6°$ **b)** $x = 41.8°$ and $x = 138.2°$ **c)** $x = 113.6°$

## Chapter 11 Extension worksheet

1. The domain of $gf(x)$ is $x \in \mathbb{R}$ and the range of $gf(x)$ is $gf(x) \geqslant 1$.

2. **a)** Domain is $x \in \mathbb{R}$ and range is $fg(x) \geqslant 1$

   **b)** Domain is $x \geq 0$ and range is $fg(x) \geqslant 2$

   **c)** Domain is $x \in \mathbb{R}, x \neq 0$ and range is $fg(x) \in \mathbb{R}, fg(x) \neq 2$

   **d)** Domain is $x \in \mathbb{R}$ and range is $fg(x) \geqslant 7$

   **e)** Domain is $x \in \mathbb{R}, x \neq 3$ and range is $fg(x) \in \mathbb{R}, fg(x) > 1$

   **f)** Domain is $x \in \mathbb{R}$ and range is $fg(x) > 2$

   **g)** Domain is $x \geq 0$ and range is $0 < fg(x) \leqslant 1$

   **h)** Domain is $x \in \mathbb{R}$ and range is $fg(x) \in \mathbb{R}$

## Chapter 12 Extension worksheet

1. **a)** $\begin{pmatrix} 1 \\ 3 \\ 7 \end{pmatrix}$
   **b)** $\begin{pmatrix} 4 \\ 2 \\ 6 \end{pmatrix}$
   **c)** $\begin{pmatrix} 8 \\ -3 \\ 1 \end{pmatrix}$
   **d)** $\begin{pmatrix} 7 \\ -2 \\ 5 \end{pmatrix}$

   **e)** $\begin{pmatrix} 4 \\ 1 \\ 6 \end{pmatrix}$
   **f)** $\begin{pmatrix} 18 \\ -5 \\ 5 \end{pmatrix}$
   **g)** $\begin{pmatrix} 3 \\ -1 \\ -1 \end{pmatrix}$
   **h)** $\begin{pmatrix} -8 \\ 1 \\ -1 \end{pmatrix}$

2. 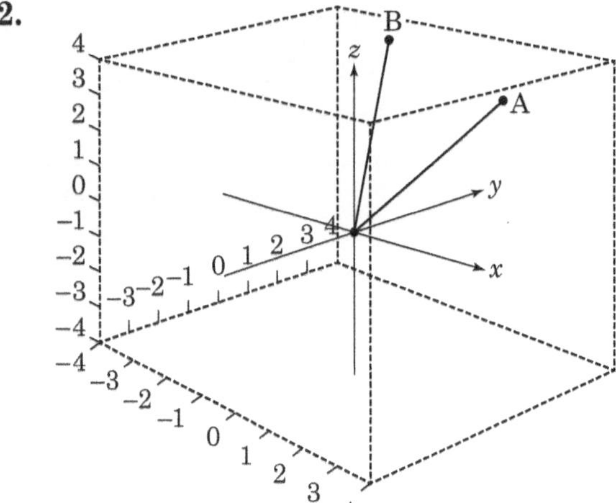


3. $\begin{pmatrix} -3 \\ 1 \\ 1 \end{pmatrix}$

4. $r = \begin{pmatrix} 2 \\ 1 \\ 3 \end{pmatrix} + \lambda \begin{pmatrix} -3 \\ 1 \\ 1 \end{pmatrix}$

5. **a)** $(-1, 2, 4)$
   **b)** $(-4, 3, 5)$

   **c)** $(5, 0, 2)$
   **d)** $(12.5, -2.5, -0.5)$

6. **a)** $\overrightarrow{OA} = \begin{pmatrix} 4 \\ 5 \\ 2 \end{pmatrix}, \overrightarrow{OB} = \begin{pmatrix} -1 \\ 6 \\ 8 \end{pmatrix}$
   **b)** $\begin{pmatrix} -5 \\ 1 \\ 6 \end{pmatrix}$

   **c)** $r = \begin{pmatrix} 4 \\ 5 \\ 2 \end{pmatrix} + \lambda \begin{pmatrix} -5 \\ 1 \\ 6 \end{pmatrix}$

## Chapter 13 Extension worksheet

1. 27.3 seconds

2. 110.5 cm

## Chapter 14 Extension worksheet

1. a) 0.0156

   b) 0.3126

   c) 0.0020

2. a) 0.0355

   b) 0.5053

   c) 0.0160

### Multiple choice revision test 1

| | | | |
|---|---|---|---|
| 1. C | 2. C | 3. C | 4. A |
| 5. D | 6. C | 7. B | 8. D |
| 9. B | 10. D | 11. C | 12. D |
| 13. B | 14. C | | |

### Multiple choice revision test 2

| | | | |
|---|---|---|---|
| 1. D | 2. D | 3. C | 4. C |
| 5. D | 6. B | 7. C | 8. C |
| 9. A | 10. A | 11. D | 12. D |
| 13. B | 14. B | | |

### Multiple choice revision test 3

| | | | |
|---|---|---|---|
| 1. B | 2. A | 3. A | 4. B |
| 5. A | 6. A | 7. C | 8. D |
| 9. D | 10. A | 11. C | 12. C |

### Multiple choice revision test 4

| | | | |
|---|---|---|---|
| 1. C | 2. C | 3. A | 4. D |
| 5. C | 6. C | 7. D | 8. D |
| 9. B | 10. A | 11. C | 12. C |

## Examination-style Paper 2 (Non-calculator) mark scheme

| Question | Answer | Marks | Partial Marks |
|---|---|---|---|
| **1.** | 76.50 | 1 | |
| **2. a)** | 8 | 1 | |
| **2. b)** | | 1 | |
| **3.** | | 3 | B1 for 7 in 'A only'<br>B1 for intersection<br>B1 for complete correct Venn diagram |
| **4.** | $216 + 18\sqrt{3}$ cm$^2$ | 3 | M1 for attempt to find height and hence of triangle<br>$(h = 3\sqrt{3},\ \text{area} = 9\sqrt{3})$<br>M1 for $6 \times 3 \times 12$ |
| **5.** | 20.42 | 2 | M1 for $700 \times 0.05$ or finding 5% by other method |
| **6.** | 418 | 1 | |
| **7.** | $\dfrac{1}{2}\mathbf{c} - \dfrac{1}{2}\mathbf{a}$ | 2 | M1 for $\overrightarrow{OB} = \mathbf{c} - \mathbf{a}$ oe |
| **8.** | $x < -\dfrac{7}{11}$ | 3 | M1 for $6x + 10$<br>M1 for $k > 11x$ |
| **9.** | $2.871 \times 10^9$ | 1 | |
| **10. a)** | $m = \dfrac{1}{2}$<br>$c = 3$ | 2 | |

| | | | |
|---|---|---|---|
| **10. b)** | 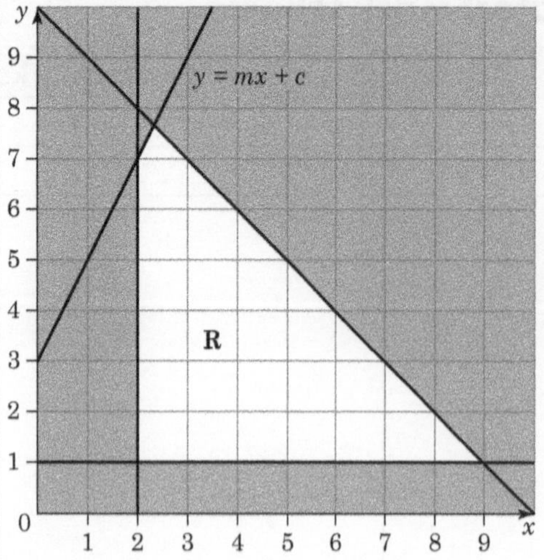 | 2 | B1 for adding vertical and horizontal lines<br>B1 (indep) for shading 'inside' region for *their* lines |
| **11.** | 2300 | 1 | |
| **12.** | $7\frac{9}{10}$ | 3 | B2 for $\frac{79}{10}$<br>or M2 for $\frac{34}{10}+\frac{45}{10}$ or $7\frac{4}{10}+\frac{5}{10}$<br>or M1 for $\frac{17}{5}+\frac{9}{2}$ or $3+4+\frac{2}{5}+\frac{1}{2}$ |
| **13.** | 2 hours 55 minutes | 1 | Accept 175 minutes |
| **14. a)** | 13 | 1 | |
| **14. b)** | $20x$ | 1 | |
| **14. c)** | 256 | 1 | |
| **15. a)** | $8\sqrt{3}$ | 2 | B1 for $2\sqrt{3}$ or $6\sqrt{3}$ |
| **15. b)** | $\frac{1+\sqrt{3}}{2}$ | 2 | M1 for $\times\frac{\sqrt{3}+1}{\sqrt{3}+1}$ |
| **16.** | 144 | 3 | M1 for $y \propto (x+1)^2$ or $y = k(x+1)^2$<br>A1 for $k = 16$ |
| **17.** | $x = 4$<br>$y = -2$ | 3 | M2 for attempt to eliminate either variable<br>A1 for one correct value |
| **18.** | $27x^3$ | 2 | B1 for 27 or $x^3$ |
| **19. a)** | $x^3(c-d)$ | 1 | |
| **19. b)** | $x = \sqrt[3]{\frac{b^2+a}{c-d}}$ | 2 | M1 for $x^3 = \frac{b^2+a}{\textit{their}\text{ bracket in (a)}}$ |
| **20. a)** | Reflection in the line $y = x$ | 2 | B1 for reflection |

| | | | |
|---|---|---|---|
| **20. b)** | | 2 | B1 for a 90° rotation drawn anywhere |
| **21. a)** | $f'(x) = 9x^2 - 8x + 2$ | 3 | M1 for evidence of differentiating, e.g. one or more terms with power reduced by 1<br>A1 two terms correct |
| **21. b)** | 22 | 1 | ft *their* (a) |
| **22. a)** | $a = \dfrac{1}{9};\ b = 1;\ c = 27$ | 3 | |
| **22. b)** | | 5 | B3 for all points plotted correctly<br>B2 for 5 or 6 points plotted correctly (ft *their* (a))<br>B1 for 3 or 4 points plotted correctly (ft *their* (a))<br>B1 (indep) for smooth curve joining *their* points<br>B1 (indep) for clear asymptotic behaviour for negative $x$ |
| **22. c)** | 2.7 | 1 | Allow 2.6–2.8<br>ft *their* (b) |
| **22. d)** | 0 | 1 | |

| | | | |
|---|---|---|---|
| **22. e)** | 4.1 | 3 | M1 for drawing suitable tangent on *their* (b)<br>M1 for evidence of gradient calculation from *their* tangent<br>Allow 4–4.2 with evidence |
| **22. f)** | 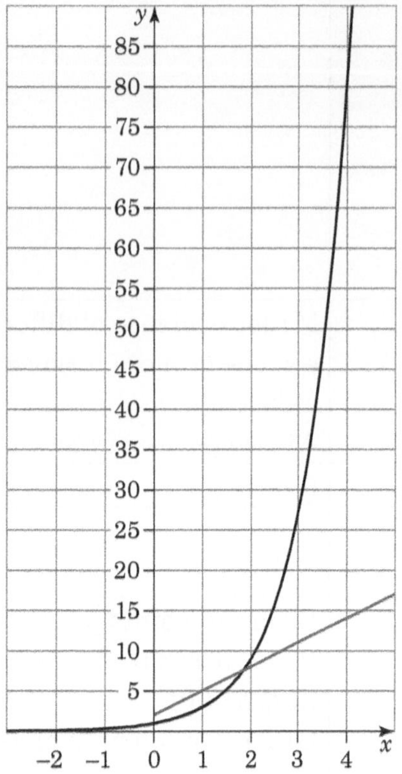 | 2 | M1 for line drawn with positive intercept and positive gradient |
| **22. g)** | 1.8 | 2 | B1 for solution ft *their* curve and line<br>Allow 1.7–1.9 |
| **23. a)** | $\frac{105}{c}$ (g) | 1 | |
| **23. b)** | $\frac{105}{c+4}$ (g) | 1 | |
| **23. c)** | $\frac{105}{c} - \frac{105}{c+4} = 0.8$<br>$105(c+4) - 105c = 0.8c(c+4)$<br>$420 = 0.8c^2 + 3.2c$<br>$c^2 + 4c - 525 = 0$ | 4 | M1 for correct equation using *their* (a) and (b)<br>M1 (indep) for clearing fractions<br>M1 (indep) for attempt to simplify<br>A1 complete correct algebra leading to answer given |
| **23. d)** | $c = 21$ | 3 | M1 for $(c \pm 25)(c \pm 21)$<br>A1 for $(c + 25)(c - 21)$<br>–25 must be rejected |
| **23. e)** | $\frac{105}{23}$ (g) | 3 | M1 for $21 + 25 = 46$ chocolates (ft *their* (d))<br>M1 for $\frac{210}{46}$ oe |
| **24. a) i)** | cyclic (quadrilateral) | 1 | |
| **24. a) ii)** | $x = 42°$; $y = 20°$; $z = 118°$ | 3 | |

| | | | |
|---|---|---|---|
| **24. a) iii)** | 90° | 1 | |
| **24. a) iv)** | 130° | 1 | |
| **24. b) i)** | Angle at $P$ equals angle at $S$<br>Angle at $Q$ equals angle at $R$<br>(Angles in same segment)<br>Angles at $X$ are equal (vertically opposite)<br>Hence corresponding angles are equal so the triangles are similar | 3 | M1 for angle $P$ = angle $S$ or angle $Q$ = angle $R$<br>M1 for opposite angles equal<br>A1 for complete correct reasoning |
| **24. b) ii)** | $6\dfrac{2}{3}$ (cm²) | 2 | M1 for area scale factor 2.25 oe |
| **25. a)** | $x = 20$; $y = 38$ | 1 | |
| **25. b) i)** | $\dfrac{1}{4}$ | 1 | |
| **25. b) ii)** | $\dfrac{40}{77}$ | 2 | B1 for 40<br>B1 for 77 |
| **25. c) i)** | $\dfrac{43}{120}$ | 1 | |
| **25. c) ii)** | $\dfrac{17}{30}$ | 2 | B1 for 17<br>B1 for 30 |

# Examination-style Paper 4 (Calculator) mark scheme

| Question | Answer | Marks | Partial Marks |
|---|---|---|---|
| 1. | 1 | 1 | |
| 2. a) | 3.2 | 1 | |
| 2. b) | $\dfrac{1}{8}$ | 1 | |
| 3. a) | $x = 9.17$<br>$y = 7.5$ | 4 | M1 for sf $\dfrac{11}{6}$ or $\dfrac{6}{11}$ (seen or implied)<br>A1 for $x$<br>M1 for finding AD using *their* sf<br>A1 for $y$ |
| 3. b) | $1024\pi$ (cm²) | 2 | M1 for sf 16 |
| 4. | $34\dfrac{7}{12} \leqslant W < 35\dfrac{5}{12}$ | 2 | M1 for 6 hours 55 minutes or<br>7 hours 5 minutes<br>Accept decimal equivalents |
| 5. | 0.343 oe | 2 | M1 for 0.7 |
| 6. | $\dfrac{75}{68}$ oe | 2 | M1 for one correct initial step e.g.<br>adding fractions on the left or<br>clearing fractions |
| 7. | 400 | 4 | M1 for $\dfrac{240 + x}{400 + x} = 0.8$<br>M1 for $240 + x = 0.8(400 + x)$<br>M1 for $0.2x = 80$ |
| 8. a) | <table><tr><td>$n$</td><td>0</td><td>1</td><td>2</td><td>3</td><td>4</td></tr><tr><td>$M$</td><td>500</td><td>650</td><td>845</td><td>1099</td><td>1428</td></tr></table> | 2 | B1 for any three correct |
| 8. b) | 6 | 1 | |
| 9. | $(-1.59, 13.18)$<br>$(1.26, 7.49)$ | 5 | M1 for $10 - 2x = 3x^2 - x + 4$<br>A1 for $3x^2 + x - 6 = 0$ oe<br>M1 (indep) for use of quadratic<br>formula on *their* three-term<br>quadratic<br>A1 for both $x$<br>A1 for both coordinates (can be given<br>in the form $x = \dots, y = \dots$) |
| 10. a) | 5.705 (minutes) | 3 | B1 for use of midpoints<br>M1 for $\dfrac{\bullet\,fx}{\bullet\,f}$ if $x$ is midpoint, or<br>consistently incorrect |
| 10. b) | | 3 | M1 for calculating frequency<br>densities<br>A1 for each bar drawn o the correct<br>height |

| 10. c) | 2.5 | 1 | |
|---|---|---|---|
| 11. a) | 7.01 (m) | 3 | M1 for attempt at arc length<br>M1 for adding *their* arc length to 4 other lengths |
| 11. b) | 3.13 (m$^2$) | 3 | M1 for attempt at sector area<br>M1 for adding *their* sector area to the area of the remaining compound shape |
| 11. c) | 12.5 (m$^3$) | 1 | ft *their* cross-section area |
| 11. d) | 25.5 (m$^2$) | 4 | M1 for area of base, back and top<br>M1 for *their* arc length $\times$ 4<br>M1 for adding *their* cross-section $\times$ 2 |
| 12. a) | (0.5, 6) | 1 | |
| 12. b) | $\sqrt{29}$ | 2 | M1 for $\sqrt{(3--2)^2+(5-7)^2}$ oe<br>Accept decimal 3 sf or better |
| 12. c) | $(y=)\ \dfrac{5}{2}x-\dfrac{43}{2}$ oe | 4 | M1 for gradient of AB $\dfrac{7-5}{-2-3}$ oe<br>M1 dep for gradient of perpendicular $p=\dfrac{-1}{their\ -\dfrac{2}{5}}$ oe<br>M1 dep on previous M1 for substituting (9, 1) into $y=their\ px+c$ |
| 13. a) | Jaipreet (by) \$31.45 | 6 | B2 for 2350<br>or M1 for $\dfrac{2000\times3.5\times5}{100}$ oe<br>M1 for $2000\times1.03^n$<br>A1 for \$2318.55<br>A1 for correct final answer |
| 13. b) | \$5097.41 | 3 | M1 for 0.88<br>M1 for $8500\times their\ 0.88^n$ |
| 14. a) | 0.393 km | 4 | M1 for $LH^2=1.4^2+1.6^2-2\times1.4\times1.6\times\cos13°$<br>M1 for attempt to simplify RHS using correct order of operations, e.g. $LH^2=4.52-4.365...$ or $LH^2=0.154...$<br>M1 (indep) for square rooting *their* $LH^2$ |
| 14. b) | 1.21 km | 4 | B1 for $WHJ=49°$<br>M1 for use of sine rule with *their* $WHJ$<br>$\dfrac{WJ}{\sin49°}=\dfrac{1.6}{\sin95°}$<br>M1 for attempt to evaluate expression for $WJ$ using *their* $WHJ$ |

| | | | |
|---|---|---|---|
| **14. c)** | 0.822 km² | 3 | M1 for either $\frac{1}{2} \times 1.4 \times 1.6 \sin13°$<br>or $\frac{1}{2} \times 1.6 \times$ *their* (b) $\sin36°$<br>M1 for adding both expressions for area |
| **14. d) i)** | 073° | 1 | |
| **14. d) ii)** | 289° | 1 | |
| **14. e)** | 1 : 40 000 | 2 | M1 for any correct intermediate ratio, e.g. 4 : 160 000 |
| **15. a) i)** | $9.90 | 2 | M1 for $3.30 \div 3 \times$ '9' |
| **15. b) ii)** | 80.2% | 2 | M1 for either *their* (a)(i) ÷ 50<br>or (50 – *their* (a)(i)) ÷ 50 |
| **15. b)** | 2.8 kg | 2 | M1 for 14.7 ÷ 21 × 4 |
| **15. c)** | 2.39 kg/$ | 3 | B1 ft for cost of juice ($4.40)<br>B1 ft for mass of juice (10.5 kg) |
| **15. d)** | $2.75 | 2 | M1 for 3.30 ÷ 1.2 |
| **16. a)** | –7 | 4 | M1 for $2(x - 2) + 3(x + 3) = -30$ oe<br>M1 for $5x + 5 = -30$ oe<br>M1 for $5x = k$ where $k \neq -5$ or $-30$ |
| **16. b) i)** | $\dfrac{25}{7}$ | 1 | Accept mixed number but not rounded decimal |
| **16. b) ii)** | $\dfrac{y + 23}{(y - 1)(y + 5)}$ | 2 | M1 for $\dfrac{4(y + 5) - 3(y - 1)}{(y - 1)(y + 5)}$<br>Accept expanded denominator |
| **16. b) iii)** | 15.64 and –0.64 | 3 | M1 for *their* (a) $= \dfrac{2}{y}$ leading to a three-term quadratic,<br>e.g. $y^2 - 15y - 10 = 0$<br>M1 for use of quadratic formula with their $a$, $b$ and $c$ |
| **16. c)** | $b = \dfrac{c}{a} - 1$ oe | 3 | M1 for $a(b + 1) = c$<br>M1 for $b + 1 = \dfrac{c}{a}$<br>or $ab = c - a$ if brackets expanded |